Sustainable Neighbourhoods for Ageing in Place

Nestor Asiamah • Hafiz T. A. Khan
Pablo Villalobos Dintrans
Mohammad Javad Koohsari
Emmanuel Mogaji • Edgar Ramos Vieira
Ruth Lowry • Henry Kofi Mensah
Editors

Sustainable Neighbourhoods for Ageing in Place

An Interdisciplinary Voice Against Global Crises

Springer

Editors
Nestor Asiamah
Division of Interdisciplinary Research and Practice School of Health and Social Care
University of Essex
Colchester, Essex, UK

Department of Gerontology and Geriatrics
Africa Centre for Epidemiology
Accra, Ghana

Pablo Villalobos Dintrans
Programa Centro Salud Pública
Facultad de Ciencias Médicas
Universidad de Santiago
Santiago, Chile

Millennium Institute for Care Research (MICARE)
Santiago, Chile

Emmanuel Mogaji
Keele Business School
Keele University
Staffordshire, UK

Ruth Lowry
School of Sports, Rehabilitation, and Exercise Sciences
University of Essex
Colchester, Essex, UK

Hafiz T. A. Khan
College of Nursing, Midwifery and Healthcare
University of West London
Brentford, UK

Mohammad Javad Koohsari
Japan Advanced Institute of Science and Technology
Nomi, Ishikawa, Japan

Waseda University
Tokorozawa, Saitama, Japan

School of Exercise and Nutrition Sciences
Deakin University
Burwood, VIC, Australia

Edgar Ramos Vieira
Department of Physical Therapy
Nicole Wertheim College of Nursing & Health Sciences
Florida International University
Miami, FL, USA

Henry Kofi Mensah
Department of Organizational and Human Resources Development
Kwame Nkrumah University of Science and Technology
Kumasi, Ghana

ISBN 978-3-031-41593-7   ISBN 978-3-031-41594-4 (eBook)
https://doi.org/10.1007/978-3-031-41594-4

This Springer imprint is published by the registered company Springer Nature Switzerland AG
The registered company address is: Gewerbestrasse 11, 6330 Cham, Switzerland

Paper in this product is recyclable.

Preface

All living things need an ecosystem to grow and thrive, and human beings particularly need convivial neighbourhoods to remain healthy and enjoy life. Neighbourhoods characterised by violence, social segregation, unequal distribution of amenities, hazardous industrial practices, natural disasters, and infectious disease outbreaks are unlikely to provide residents with opportunities for optimal health, life enjoyment, and healthy longevity. In such neighbourhoods, residents look for opportunities to leave for more homely communities, live in constant fear, or experience a faster decline in physical and mental health over the life course. Consequently, residents are more likely to live in care homes or places less satisfactory and supportive of a healthy enjoyable life. In recent years, climate change events (e.g., extreme weather, flooding, and tsunamis), extreme or radical industrialisation, the outbreak of infectious diseases, and systemic violence, hereby collectively referred to as global crises, are being felt worldwide and are expected to become more commonplace. These crises may, thus, obliterate neighbourhood psychosocial factors (e.g., peace, safety, and social cohesion) and exterminate built environment resources (e.g., service infrastructure and parks) that facilitate social engagement and other pro-health behaviours. The crises can, thus, make community life more daunting and increase the amount of resilience needed by residents to age well in their neighbourhoods.

This book aims to improve stakeholders' understanding of what a sustainable neighbourhood is in the context of crises by presenting problems and solutions from different countries and disciplines. This book: (1) provides an analysis of how the crises affect neighbourhood attributes (e.g., services, residential density, and amenities) and the ability of residents to use them to maintain health while living in their preferred neighbourhoods, and (2) suggests potential interventions for enabling residents to utilise these attributes for health while living at home in contexts experiencing the crises. This book is the first to consider the four crises as health and social threats to healthy longevity from a sustainability perspective. It presents a voice of scholars and practitioners from various disciplines comprising public health, health care, architecture, engineering, human resources development, information technology, and finance.

Colchester, Essex, UK	Nestor Asiamah
Brentford, UK	Hafiz T. A. Khan
Santiago, Chile	Pablo Villalobos Dintrans
Nomi, Ishikawa, Japan	Mohammad Javad Koohsari
Staffordshire, UK	Emmanuel Mogaji
Miami, FL, USA	Edgar Ramos Vieira
Colchester, Essex, UK	Ruth Lowry
Kumasi, Ghana	Henry Kofi Mensah

Contents

About the Editors

Nestor Asiamah, PhD, is a Lecturer at the School of Health and Social Care, University of Essex, and Executive Director at the Africa Centre for Epidemiology. His research interests include gerontology and health promotion. He serves as an editorial board member of several journals, including *BMC Public Health* (Springer) and *PLoS ONE*. He publishes with his global research network an average of 10 peer-reviewed scientific papers a year.

Hafiz T. A. Khan, PhD, is a Professor of Public Health and Statistics at the University of West London, UK. He is also an Associate Professorial Fellow at the Oxford Institute of Population Ageing, University of Oxford. Prof Khan has published several articles in top-tier journals and has played editorial roles in Health Sciences journals.

Pablo Villalobos Dintrans, DrPH, is an independent consultant working on public policy and public health. He holds a BA and MA in Economics and Public Policy from the Pontificia Universidad Católica de Chile and an MA in Economics from Boston University. He completed his doctoral training at Harvard University. Pablo is an interdisciplinary researcher whose research interests cut across ageing, public health, public policy, and economics. He currently collaborates with the Millennium Institute for Care Research (MICARE), the African Health and Aging Research Center (AHARC), and the World Health Organization's Global Network on Long-Term Care (GNLTC).

Mohammad Javad Koohsari, PhD, is an Associate Professor at the Japan Advanced Institute of Science and Technology, Japan. His research focuses on how the built environment can contribute to population health. In 2020 and 2021, he was recognised in the top 2% of most influential researchers worldwide across all scientific disciplines (Stanford University and Elsevier).

Emmanuel Mogaji, PhD, is an Associate Professor in Marketing at Keele University, United Kingdom. His research interests include advertising, especially in higher education, and the application of artificial intelligence. Emmanuel is highly published, has filled several editorial roles, and edited books published by reputable publishers such as Springer.

Edgar Ramos Vieira, PhD, is an Associate Professor in Physiotherapy at the Nicole Wertheim College of Nursing and Health Sciences, Florida International University, USA. He is also the Editor-in-Chief of *Physical & Occupational Therapy in Geriatrics*, which is hosted by Taylor & Francis. Edgar is recognised by Clarivate as a highly cited author.

Ruth Lowry, PhD, is a Reader (Associate Professor) in Sports Psychology and the Director of Research at the School of Sports, Rehabilitation, and Exercise Sciences, University of Essex, UK. Ruth's research focuses on cohesive neighbourhoods supporting exercise and social participation.

Henry Kofi Mensah, PhD, is an Associate Professor at the Department of Human Resources and Organizational Development, KNUST. His research interests include green HR management, sustainability, occupational health, and organisational behaviour. Henry is also the Director of Doctoral Research Programmes at KNUST School of Business, Kumasi-Ghana.

About the Contributors

Evelyn N. Alvarez, PhD, MPH, is an Environmental Health Scientist and Professor at California State University, Los Angeles. Her research agenda largely focuses on environmental smartphone applications and wearable technology to promote environmental justice. Her research interests also include examining underrepresented narratives in the climate change dialogue and demystifying sustainable living.

Emelia Danquah, PhD, is an Organisational Psychologist and a Senior Research Fellow in Human Resources Development at Koforidua Technical University. Her research interest focuses on organisational psychology, development of an ageing workforce, and personnel grooming. Emelia has a master's degree from the University of Ulster (UK) and a PhD from Accra Institute of Technology and the Open University of Malaysia. She is the Human Resources Director at the Africa Center for Epidemiology.

Paulina Del Solar, PhD, is an Occupational Therapist and a Professor at the School of Occupational Therapy, Faculty of Medicine, Universidad de Los Andes, Santiago, Chile. Her research interests are age-friendly cities, built environments, walkability, older people, aged, physical activity, functionality, and activity of daily living.

Antonia Echeverría, PhD, is an Occupational Therapist and extraordinary Assistant Professor at the School of Occupational Therapy, Universidad de Los Andes, Santiago, Chile. She is the Director of the FONIS SA20I0097 project, funded by the National Research and Development Agency, Chile. She is also a member of the Millennium Institute for Care Research (MICARE), Santiago, Chile. Her research interests are healthy ageing, functionality, older people, dependence, activity of daily living (ADL), health prevention and promotion.

Rodrigo Fernández, PhD, is a Civil Engineer (Universidad de Los Andes) with MSc and PhD degrees in Transport Studies from the University of London. He is also a full Professor at the Faculty of Engineering and Applied Sciences, Universidad de Los Andes, Santiago, Chile. His research interests are traffic flow theory, traffic engineering, and public transport.

Jamie P. Halsall, PhD, is a Reader in Social Sciences at the Department of Behavioural and Social Sciences, University of Huddersfield, UK. Jamie's research interests are encompassed within three distinct areas: (1) civil society, social enterprise, and public policy; (2) sustainability and globalisation, and (3) coaching and mentorship.

Amar Kanekar, PhD, is a Professor and Graduate Program Coordinator for Health Education and Health Promotion at the University of Arkansas at Little Rock. His research areas of interest focus on adolescent health, measurement in health education, global health, online and hybrid pedagogy, and health behaviour interventions.

Sadasivam Karuppannan, PhD, is a Senior Lecturer in Urban and Regional Planning at the University of South Australia. His research is on housing markets, housing for the ageing population, planning for sustainable cities and household energy consumption. He has published extensively on ageing, housing, city planning, sustainability, and ecological footprints.

Kyriakos Kouveliotis, PhD, FRSA is a Professor and currently the Provost & Chief Academic Officer of the Berlin School of Business and Innovation in Germany. He also serves as a Professor at the Uninettuno University of Rome, Italy, and President at Atheneum Liberal Studies. In the last 20 years, he has guided and taught in many universities and educational organisations globally. He is an expert in developing new curricula, programs, syllabi, and on e-learning and virtual learning. Professor Kouveliotis has published 12 books and dozens of original scientific articles. Recently, he received a fellowship in the Royal Society of Arts and created in the Hague his own not-for-profit organisation, the "Global Degree Foundation".

Özge Kutlu graduated from Social Work and Social Policy Master's Program in 2022. Kutlu, who worked as a Social Worker at Burdur Provincial Health Directorate between 2019 and 2023, has been working as a Lecturer in Burdur Mehmet Akif Ersoy University Elderly Care Program since February 2023. She works on the provision of rights-based social work practices for the elderly and their families.

Liping Mei, PhD, is an Associate Professor at Beijing Youth Politics College. Liping is a member of the Chinese Psychological Society and the Vice President of China Pharmaceutical Culture Society Health Science Popularization Branch. Her research interests are encompassed within two areas: (1) ageing and elderly care, and (2) comparative study of social welfare policies.

Thomas W. Nesser, PhD, is a Professor in the Department of Kinesiology, Recreation, and Sport at Indiana State University, where his primary focus is the development and teaching of courses in human performance, and strength and conditioning. His PhD is in Kinesiology from the University of Minnesota.

Whitney Nesser, PhD, is an Associate Professor in the Department of Applied Clinical and Educational Sciences at Indiana State University. Her primary focus is on research and teaching about quality of life, gerontology, and intergenerationology. Her PhD is in Health Education and Promotion from the University of Alabama at Birmingham.

Mehmet Öçal, PhD, completed his doctorate education at Labor Economics and Industrial Relations, Pamukkale University, in 2022. He has been working as an Assistant Professor in the social work department of Burdur Mehmet Akif Ersoy University. In 2022, he was deemed worthy of the Outstanding Achievement Award from Türkiye Council of Higher Education.

Charles Prempeh, PhD, is a Research Fellow at the Centre for Cultural and African Studies, Kwame Nkrumah University of Science and Technology, Kumasi, Ghana. He had his PhD degree in Theology and Religious Studies from the University of Cambridge, United Kingdom, in 2021.

Sarra Sghaier, PhD, is a Lecturer and Researcher in Marketing. Her research is interdisciplinary and concentrates on marketing, psychology, sociology, anthropology, and gerontology. She received her first and second degrees as well as PhD in Marketing from the Faculty of Economics Sciences and Management, University of Tunis, Tunisia. Her PhD was on the impact of subjective age on innovativeness.

Kalim Siddiqui, PhD is an economist, specialising in International Political Economy, Development Economics, and Economic Policy. He is a Senior Lecturer in the Department of Accounting, Finance and Economics, University of Huddersfield. He has a wide range of research interests, and they can be classified into four broad domains: (1) political economy, (2) development economics, (3) economic policy, and (4) international trade.

Alpana Sivam, PhD, is a Senior Lecturer in Urban and Regional Planning at the University of South Australia. Her research focus is on housing and creating sustainable, climate-resilient, and age-friendly neighbourhoods through interdisciplinary collaboration and innovative design and policy solutions. Her scholarly outputs have been documented in over 100 peer-reviewed publications.

Michael Snowden, PhD, is a Senior Lecturer of Public Health at the University of Huddersfield, UK. He has a special interest in social enterprise, coaching, and mentoring, and how they can be used to promote equality and inclusivity in learning and society. During Mike's long and established career in Higher Education, he has contributed to, led, and developed numerous funded innovative projects, distinct pedagogical innovations, and conceptual models that influence and enhance learning and practice. His expertise is widely recognised as instrumental in developing curricula that support learning enhancement. Mike has published his work widely and is a regular speaker at national and international conferences.

Ali Soltani, PhD, is a Senior Researcher in Computational Urban Planning and Policy, specialising in land use and transportation. He holds degrees in Engineering, Urban Planning, and Data Science. He has published research on urban analytics, property data, and transportation. He has won awards and served as an editor for several journals.

James Stockton is a Post-graduate Researcher based at the School of Human and Health Sciences, University of Huddersfield. His research interests cover four main areas: health impacts of loneliness isolation and alienation, global development, pedagogy of higher education, and statistical research methods.

List of Figures

List of Tables

Chapter 1
Sustainable Neighbourhoods for Ageing in Place in a World of Crises: An Introduction

Nestor Asiamah, Hafiz T. A. Khan, Pablo Villalobos Dintrans, Mohammad Javad Koohsari, Emmanuel Mogaji, Edgar Ramos Vieira, and Henry Kofi Mensah

1.1 Introduction

Since the first person-environment fit model was published in 1951 to date (Asiamah et al., 2023; Wahl & Gerstorf, 2020), research on the role of the environment in health has gained significant momentum. This progress has improved the understanding of how social and physical environments influence health across the lifespan. The physical environment, characterized mainly by built environment factors (e.g. mixed land use, residential density, and street connectivity), determines whether residents can utilize neighbourhood services and perform health-seeking

N. Asiamah (✉)
Division of Interdisciplinary Research and Practice, School of Health and Social Care, University of Essex, Colchester, Essex, UK

Department of Gerontology and Geriatrics, Africa Centre for Epidemiology, Accra, Ghana
e-mail: n.asiamah@essex.ac.uk; nestor.asiamah@ace-gh.org

H. T. A. Khan
College of Nursing, Midwifery and Healthcare, University of West London, Brentford, UK
e-mail: hafiz.khan@uwl.ac.uk

P. Villalobos Dintrans
Programa Centro Salud Pública, Facultad de Ciencias Médicas, Universidad de Santiago, Santiago, Chile

Millennium Institute for Care Research (MICARE), Santiago, Chile
e-mail: pvillalobos.d@gmail.com

M. J. Koohsari
Japan Advanced Institute of Science and Technology, Nomi, Ishikawa, Japan

Faculty of Sport Sciences, Waseda University, Saitama, Japan

School of Exercise and Nutrition Sciences, Deakin University, Burwood, VIC, Australia
e-mail: koohsari@jaist.ac.jp

N. Asiamah et al. (eds.), *Sustainable Neighbourhoods for Ageing in Place*,
https://doi.org/10.1007/978-3-031-41594-4_1

behaviours such as physical activity (PA) as they age. A growing body of studies suggests that PA is higher in more walkable neighbourhoods with attributes such as higher residential density, street connectivity, and mixed land use (Gan et al., 2022; Owen et al., 2007; Roy et al., 2021; Sundquist et al., 2011). The evidence further suggests that such contexts better support health and longevity (Marquet et al., 2017).

The role of the social environment, including social networks, in health-seeking behaviours and health is also strongly recognized in the literature (Asiamah et al., 2023; Friedman & Kennedy, 2021; Gan et al., 2022; Wahl & Gerstorf, 2018). For instance, the ability of the individual to maintain PA and health over the life course depends on social support from social networks (Asiamah et al., 2023; Atchley, 1989). Other important facets of the social environment include psychosocial factors (e.g. safety, trust, and social cohesion) that facilitate social and physical activities (Asiamah et al., 2023; Wahl & Gerstorf, 2018); people living in environments without these factors can hardly maintain healthy behaviours and health across the lifespan. Yet, the contributions of the social and physical environments to health depend on personal factors (Asiamah et al., 2023; Wahl & Gerstorf, 2018).

Personal factors (e.g. functional ability and income) influence the utilization of the physical and social environment (Asiamah et al., 2023). For instance, frail individuals are less likely to utilize neighbourhood attributes (e.g. services) through walking and social participation. Thus, health and the ability to maintain it in the ageing process are an outcome of the interplay between the physical and social environment as well as personal factors. Researchers recognize that climate change, violence, infectious diseases, and extreme industrialization, hereby referred to as crises, either obliterate physical and social environment factors necessary for healthy living or make it difficult for people to use neighbourhood resources (e.g. social support and services) to perform healthy behaviours over the life course. These crises are being experienced on a global scale (Asiamah, 2022; Benevolenza & DeRigne, 2019) and can, therefore, be expected to affect the ability of individuals to maintain health in their homes and neighbourhoods as they age. Thus, "ageing in place" is less achievable amidst these crises.

E. Mogaji
Keele Business School, Keele University, Staffordshire, UK
e-mail: e.mogaji@keele.ac.uk

E. R. Vieira
Florida International University, Department of Physical Therapy, Nicole Wertheim College of Nursing & Health Sciences, Miami, FL, USA
e-mail: evieira@fiu.edu

H. K. Mensah
Department of Organizational and Human Resources Development, Kwame Nkrumah University of Science and Technology, Kumasi, Ghana
e-mail: hkmensah@knust.edu.gh

1.2 Aim and Objectives

"Ageing in place" concerns maintaining a healthy life in one's home and neighbourhood rather than needing to transition into a residential care facility in later life (Pani-Harreman et al., 2020). For people to age in place, they must maintain health or avoid health problems that may require their admission to a long-term care facility. People ageing in place remain in their preferred homes and neighbourhoods and are not compelled by unfavourable life conditions (e.g. war or infectious disease outbreak) to abandon their homes for less satisfactory places. From this viewpoint, it is understandable that the crises would threaten ageing in place by discouraging healthy behaviours relevant to optimal health or compelling individuals to leave their homes for unfamiliar places. If so, people not ageing in place may miss out on neighbourhood resources and the desired influence of the environment on healthy longevity. Given this possibility, we posit that neighbourhoods need to be made as resilient as possible to support ageing in place, particularly during crises.

Over the past decades, stakeholders have agreed that the effective design of neighbourhoods to support health and ageing in place is a product of multidisciplinary interventions and dialogue (Rudnicka et al., 2020; Yazdanpanahi & Hussein, 2021). This agreement is based on the principle of co-design, which emphasizes a need for experts in different fields (e.g. public health, healthcare, architecture, engineering, human resources development, information technology, and finance) to collaboratively develop and roll out models of the healthiest environments. Co-creating neighbourhoods can better meet the needs of all segments of the population sustainably. Yet, multidisciplinary research and dialogues on how to use neighbourhoods to maintain health for ageing in place during the crises are limited in the literature.

This book provides and contextualizes such multidisciplinary dialogue in the form of conceptual frameworks and empirical evidence. The world scene is changing rapidly partly due to the worsening crises and population ageing. With this book, scholars in different disciplines provide a nuanced analysis of potential approaches to the design of sustainable neighbourhoods or contexts that support ageing in place despite the present and future crises. The aim of the chapters comprising this book is to improve the understanding of what a sustainable neighbourhood in the context of the crisis is by presenting problems, solutions, and challenges coming from different countries and disciplines. This is the first time the four crises (i.e., climate change events, the outbreak of infectious diseases, radical industrialization, and violence) have been considered concurrently in analysing the ideal place for ageing from a sustainability perspective.

1.3 Focus and Contents of the Book

The chapters are indicative contents comprising theoretical models and empirical evidence that provide an understanding of the impacts of the crises on ageing in place as well as pathways in redesigning neighbourhoods to facilitate ageing in place in contexts experiencing the crises. They represent a significant area of research contributed by scholars and practitioners in a spectrum of topics in environmental science and gerontology (i.e. environmental gerontology).

The task of compiling this book implied a three-tier challenge: first, providing a global perspective on the debate about sustainable neighbourhoods, acknowledging its condition of global challenge, and consequently, the need for a broad discussion and several points of view; second, including proposals and points of view coming from different disciplines to recognize the need for a multidisciplinary approach in the debate; third, to meet high standards of academic research and provide a useful material for scholars, policymakers, and general public. To achieve these goals, an international call for papers was followed by a double-blind peer review, a process that ended up in the 11 chapters included in this book. This book includes contributions from 28 authors from 6 continents (i.e. Africa, Asia, Europe, North America, Oceania, and South America), from at least 9 countries (i.e. Ghana, Chile, China, Germany, Japan, Tunisia, Turkey, the United Kingdom, and the United States). The range of disciplines covered in this book include public health, architecture, civil engineering, built environment, finance, planning, marketing, human resources development, and sociology.

This first chapter (Chap. 1) provides an overview of the book and introduces the themes and contents of the other chapters. The remaining chapters are grouped into two themes; the first theme comprises five chapters focused on the adverse effects of the crises on neighbourhoods, whereas the second theme (consisting of six chapters) elucidates potential interventions to the crises to support ageing in place.

The discussion for Theme I starts with Chap. 2 where Asiamah et al. (2024c) propose an integrated theoretical model for understanding the potential impacts of the crises on human ecosystems and ageing in place. The authors recognize the crises as a threat to neighbourhood cohesion and population health, arguing that they can make coping with psychological, social, and physical challenges experienced in ageing more daunting. With what promises to serve as a theory of diminishing life space, the authors reason that residents experiencing the crises are facing a diminishing life space and a change in the 'social status' of their communities, which has undesirable implications for health-seeking behaviours and ageing in place. A noteworthy facet of the proposed framework is six postulates encompassing both consequences and potential interventions to a diminishing life space due to the crises.

Nesser and Nesser (2024) in Chap. 3 provide a nuanced intergenerational analysis of the impacts of the crises on older adults' functioning and health, highlighting age differences in the impact of the crises on functioning and health proxied with the Word Health Organization's domains (e.g. physical health, psychological health, level of independence, social relationships, and environment) of a quality-of-life measure. The authors emphasized the role of resilience and autonomy in ageing in place during the crises and how these important personal qualities can be affected by the crises. More importantly, research gaps and directions as well as recommendations for improving autonomy and resilience, especially during crises, are discussed.

In Chap. 4, Echeverría et al. (2024) utilize a multidisciplinary perspective to describe functioning and social participation during one or more of the crises. The authors highlight the role of age-friendly cities in health-seeking transportation behaviour and subsequently espouse how different public transport routes impact older adults' functioning. To substantiate their argument, the authors describe a state-of-the-art in Chile, providing real-life events signifying a link between public transport routes, mobility, and social participation among older adults. Their provision of this special case of Chile set the foundation for understanding their analysis of studies on related ideas from different countries. This analysis premises their thoughts around an international context and unfolds some implications for policy and stakeholder action.

Öçal and Kutlu (2024) in Chap. 5 utilize data from publicly available databases in Europe to analyse standard indicators of abuse and neighbourhood safety in Europe and Turkey to set the basis for discussing implications of neighbourhood safety and abuse for ageing in place. The authors further compare the data from Europe and Turkey to uncover key differences in the evidence as well as similarities in abuse and neighbourhood safety between Turkey and the rest of Europe. Öçal and Kutlu reveal with their analysis that the abuse of older adults and poor safety are influenced by socio-economic status. Thus, older adults with low socio-economic status are more likely to experience abuse and report low safety in their neighbourhoods. They also report that older adults living in neighbourhoods with more interconnected streets and sidewalks reported fewer barriers to independent movement, which suggests that neighbourhood attributes can influence perceived individual mobility.

Chapter 6 by Asiamah et al. (2024b) builds on the safety ideas of Öçal and Kutlu (2024) by providing a theoretical heuristic concerning the evolution of neighbourhood services over time. With this framework, the authors argue that extreme industrialization as a worsening phenomenon can terminate the evolution of new neighbourhoods into highly walkable contexts where health-seeking behaviours are encouraged and sustained for optimal health and ageing in place. Accompanying this idea is the concept of 'Neighbourhood Service Mix' and how it makes a downturn due to extreme industrialization within the neighbourhood development continuum. The authors proposed different services within the Neighbourhood Service Mix and the respective impacts of these services at four stages (i.e. budding, pre-community, community, and extreme industry) of the continuum. Finally, the chapter unfolds strategies for pre-empting the stage of 'extreme industry' that include maximizing neighbourhood safety by implementing policies to control industrial activities in communities.

Opening Theme II is Chap. 7 by Sivam et al. (2024) who utilize data from Australia to demonstrate the role of built environment attributes in access to amenities. Before their comprehensive analysis of the data, the authors provide a theoretical model demonstrating the roles of the built environment and its constituent amenities in ageing in place. Key amenities recognized in the literature are tabulated as the focus of the statistical analysis and subsequently discussed. More noticeable is the authors' analysis of the correlation between the density of the older adult population and accessibility to amenities in various suburbs of Australia, which demonstrates the role of neighbourhood design and planning in access to services and other neighbourhood attributes among older adults. Outstanding in the analysis is a projection of older adults' access to amenities in Australia for the next 30 years, which may provide cues for planning within counties and districts.

In Chap. 8, Prempeh (2024) questions society's biased view of older adults and old age, arguing that the worth and dignity of people (e.g. older adults) should not be measured with their economic worth and contribution to society, which is demonstrated through participation in the labour market. Drawing on religious beliefs and personal experiences in Ghana, Prempeh recognized older adults as an important segment of the general population characterized by rich life experience, wisdom, and piety. While older adults support younger generations with these qualities, their marginalization in society might undermine the age-friendliness of their communities; thus, even the most age-friendly contexts with ageist stereotypes may not encourage social inclusion among older adults and may, therefore, not contribute to ageing in place. Discrimination and undervaluation of older adults in society may more strongly discourage social inclusion and health-seeking behaviours amidst crises.

In Chap. 9, Asiamah et al. (2024a) formulate a theoretical framework that provides an understanding of the role of assistive technologies in ageing in place in contexts experiencing crises. Through the lens of a theoretical framework and the *Technology Demands-Resources Concept*, the authors identify different human development pathways (e.g. training, formal education, observation, and experimentation) to enable older adults to effectively use assistive technologies for ageing in place. Drawing on relevant life course development theories and previous research, the authors identify the major barriers to the use of assistive technologies among older adults or people experiencing a decline in physical and mental health over the life course. Finally, Asiamah and colleagues suggest recommendations and implications for human development. These recommendations and the basic imports of the chapter are summarized in six postulates, which emphasize the barriers to technology use among older adults as well as recommended stakeholder actions.

In Chap. 10, Alvarez (2024) identifies and analyses the various components of ageing in place (e.g. social participation, mental health, agency, access to healthcare, exercise and mobility, and technology use) affected by the coronavirus disease 2019 (COVID-19). The author identified social participation as one of the most important dimensions of ageing in place affected by the crises and tabulated potential interventions through which stakeholders can mitigate the adverse impacts of COVID-19 or infectious diseases on the various dimensions. City planning, creating a feeling of autonomy, maintenance of exercise, and the creation of an emergency preparedness plan are noteworthy interventions reported. Outstanding in this chapter is the '15-minute city' concept, which refers to a city or community where services and other neighbourhood resources can be reached within a 15-minute walk away from home. This concept emphasizes a need for compact cities or contexts with a higher population density as presented by Sivam et al. (2024) in Chap. 7.

Chapter 11 by Halsall et al. (2024) provides an interdisciplinary analysis of the demographic change in China and India from the perspective of industrial and economic revolution. First, Halsall and colleagues review the respective economic transportations in China and India that form the basis of ageing and demographic transition in the two countries and thereafter compare the two countries, highlighting their similarities and differences. The authors reveal that the rate of population ageing in China, compared to India, is higher, which means that the two countries face different burdens in meeting the needs of the older adult population, especially in places experiencing crises. Population ageing and the extent to which the crises would impact ageing in place have been influenced by policies rolled out in both countries. A policy that has strongly affected ageing in China is the 'one-child-to-one-couple' policy, which has been amended over the past years. Finally, the chapter reports implications for ageing in place and the welfare of the elderly, considering the crises.

In Chap. 12, Asiamah (2024) introduced the concept of 'sustainable ageing' in the context of ageing in place during the crises. This concept is operationalized through a review of key healthy ageing initiatives, including the Decade of Healthy Ageing (2016–2020) campaign. Through the lens of a theoretical framework, the author operationally defines sustainable ageing and its dominant dimensions (e.g. a preventive approach to ageing, elimination of inequity in health, infusion of technologies in ageing, and the creation of age-friendly cities). According to the chapter, sustainable ageing is not a replacement for existing terminologies such as 'successful ageing' and 'healthy ageing' but serves as a flagship concept describing different pathways to enabling people to age in good health amidst crises. Thus, the empowerment of older adults and populations to age well in place despite the crises is the heart of sustainable ageing. The chapter concludes with the concept of 'co-design' which emphasizes a need for all stakeholders to be involved in planning and implementing programmes for ageing in place in contexts experiencing the crises.

1.4 Significance, Limitations, and Future Research

Given the adverse impacts of the crises on health and ageing, researchers have expounded implications of the crises for ageing and proffered pathways toward alleviating or avoiding these impacts on the general population (Asiamah, 2022; Gamble et al., 2013; Leyva et al., 2017; Sadruddin & Inhorn, 2020; Strong et al., 2015). These contributions, nevertheless, have been focused on one or a couple of the crises without an emphasis on ageing in place. Thus, this edited book was the first to concurrently address these four crises regarding their impacts on older adults and their ageing in place, thereby using evidence and theoretical models to demonstrate the individual and joint impacts of the crises on ageing in place. As a unique contribution to the literature, multidisciplinary efforts enabling older adults to age successfully in place amidst crises are delineated with evidence or theoretical models. This book's novelty is characterized by a multidisciplinary voice that lends to the plausibility of its concept of neighbourhood 'co-design', which refers to the involvement of actors from relevant fields (e.g. engineering, architecture, health, human resources development, information technology, sociology, and planning) in planning or redesigning neighbourhoods resilient against the crises.

This book precludes important narratives needed to assess the enormity of the impacts of the crises on older adults and ageing in place. One such narrative is an integrative review of case studies on how each of the crises has disorientated older adults around the world. This review is necessitated by the prevalence of the crises on a global case, with examples being the war in Ukraine (Barchielli et al., 2022) and climate change events such as flooding, thermal activities (e.g. volcanoes), and extreme weather (Asiamah, 2022; Gamble et al., 2013). Limiting the book's multidisciplinary voice is the absence of chapters and sections presented in important disciplines including financial accounting and building technology. Expertise from these disciplines would play a major role in planning and designing neighbourhoods resilient against crises. Expertise in accounting and finance, for example, is needed to estimate and project the cost of interventions aimed at improving the living conditions of residents in communities experiencing crises.

Given the above limitations, researchers are encouraged to provide a narrative on the foregoing case studies, enabling stakeholders to assess and appreciate the gravity of the crises on older adults and their ageing in place. Though the crises are globally felt, their impacts on societies with different cultural orientations may differ. As such, any compilation of such case studies may demonstrate variations in the impacts of the crises across contexts and time. Similar works that include all the relevant disciplines or compensate for the exclusion of some disciplines in the current book are needed to present a holistic multidisciplinary voice or create a complete narrative of the impacts of the crises on ageing in place and their respective remedies.

1.5 Conclusion

Neighbourhoods, as human ecosystems, play an important role in health, economic growth, and healthy longevity. Yet, its attributes (e.g. safety, social cohesion, and trust) that facilitate healthy ageing are increasingly susceptible to crises. This edited work is a compilation of models and empirical evidence regarding the impacts of the crises on the neighbourhood and requisite measures that can be taken to enable older adults to utilize neighbourhoods for ageing place despite the crises. The contents of this book are a multidisciplinary voice highlighting relevant research and policy agenda in different disciplines such as engineering, architecture, health, information technology, and planning. In conclusion, neighbourhoods that are resilient against crises and facilitate healthy ageing are an outcome of policy interventions in which different actors (e.g. politicians, civil society, and researchers) are involved in city or town planning, design, or redesign while recognizing the current and future impacts of the crises on neighbourhoods. To provide enough evidence for interventions, research ought to span an exhaustive spectrum of the above disciplines.

References

Alvarez, E. N. (2024). Infectious diseases and healthy ageing: Making the case for a 15-minute city. In N. Asiamah, H. T. A. Khan, P. V. Dintrans, M. J. Koohsari, E. Mogaji, E. R. Vieira, R. Lowry, & H. K. Mensah (Eds.), *Sustainable neighbourhoods for ageing in place: An interdisciplinary voice against global crises*. Palgrave.

Asiamah, N. (2022). Walkable urban neighborhoods: The adverse effects of industrialization and climate change in developing countries. *Sustainable Urbanism in Developing Countries, March*, 247–262. https://doi.org/10.1201/9781003131922-16

Asiamah, N. (2024). "Sustainable ageing" in a world of crises. In N. Asiamah, H. T. A. Khan, P. V. Dintrans, M. J. Koohsari, E. Mogaji, E. R. Vieira, R. Lowry, & H. K. Mensah (Eds.), *Sustainable neighbourhoods for ageing in place: An interdisciplinary voice against global crises*. Palgrave.

Asiamah, N., Bateman, A., Hjorth, P., Khan, H. T. A., & Danquah, E. (2023). Socially active neighborhoods: Construct operationalization for aging in place, health promotion and psychometric testing. *Health Promotion International, 38*(1), 1–10.

Asiamah, N., Danquah, E., Sghaier, S., Mensah, H. K., & Kouveliotis, K. (2024a). Assistive technologies for ageing in place: A theoretical proposition of human development postulates. In N. Asiamah, H. T. A. Khan, P. V. Dintrans, M. J. Koohsari, E. Mogaji, E. R. Vieira, R. Lowry, & H. K. Mensah (Eds.), *Sustainable neighbourhoods for ageing in place: An interdisciplinary voice against global crises*. Palgrave.

Asiamah, N., Kanekar, A., Khan, H. T. A., & Dintrans, P. V. (2024b). Neighbourhood services and ageing in place: An extreme industrialisation perspective. In N. Asiamah, H. T. A. Khan, P. V. Dintrans, M. J. Koohsari, E. Mogaji, E. R. Vieira, R. Lowry, & H. K. Mensah (Eds.), *Sustainable neighbourhoods for ageing in place: An interdisciplinary voice against global crises*. Palgrave.

Asiamah, N., Koohsari, M. J., & Lowry, R. (2024c). Ageing in place: The present and future social and health threats. In N. Asiamah, H. T. A. Khan, P. V. Dintrans, M. J. Koohsari, E. Mogaji, E. R. Vieira, R. Lowry, & H. K. Mensah (Eds.), *Sustainable neighbourhoods for ageing in place: An interdisciplinary voice against global crises*. Palgrave.

Atchley, R. C. (1989). A continuity theory of normal aging. *Gerontologist, 29*(2), 183–190. https://doi.org/10.1093/geront/29.2.183

Barchielli, B., Cricenti, C., Gallè, F., Sabella, E. A., Liguori, F., Da Molin, G., Liguori, G., Orsi, G. B., Giannini, A. M., Ferracuti, S., & Napoli, C. (2022). Climate changes, natural resources depletion, COVID-19 pandemic, and Russian-Ukrainian war: What is the impact on habits change and mental health? *International Journal of Environmental Research and Public Health, 19*(19). https://doi.org/10.3390/ijerph191911929

Benevolenza, M. A., & DeRigne, L. A. (2019). The impact of climate change and natural disasters on vulnerable populations: A systematic review of literature. *Journal of Human Behavior in the Social Environment, 29*(2), 266–281. https://doi.org/10.1080/10911359.2018.1527739

Echeverría, R. T. A., del Solar, P. D., & Fernández, R. A. (2024). Older people functionality and community participation: An interdisciplinary health-transport approach for age-friendly cities. In N. Asiamah, H. T. A. Khan, P. V. Dintrans, M. J. Koohsari, E. Mogaji, E. R. Vieira, R. Lowry, & H. K. Mensah (Eds.), *Sustainable neighbourhoods for ageing in place: An interdisciplinary voice against global crises*. Palgrave.

Friedman, E. M., & Kennedy, D. P. (2021). Typologies of dementia caregiver support networks: A pilot study. *The Gerontologist, XX*(Xx), 1–10. https://doi.org/10.1093/geront/gnab013

Gamble, J. L., Hurley, B. J., Schultz, P. A., Jaglom, W. S., Krishnan, N., & Harris, M. (2013). Climate change and older Americans: State of the science. *Environmental Health Perspectives, 121*(1), 15–22. https://doi.org/10.1289/ehp.1205223

Gan, D. R. Y., Mahmood, A., Routhier, F., & Mortenson, W. B. (2022). Walk/wheelability: An inclusive instrument pair for participatory age-friendly research and practice. *Gerontologist, 62*(1), E39–E47. https://doi.org/10.1093/geront/gnab079

Halsall, J. P., Mei, L., Siddiqui, K., Snowden, M., & Stockton, J. (2024). Demographic changes and ageing in China and India: A public policy perspective. In N. Asiamah, H. T. A. Khan, P. V. Dintrans, M. J. Koohsari, E. Mogaji, E. R. Vieira, R. Lowry, & H. K. Mensah (Eds.), *Sustainable neighbourhoods for ageing in place: An interdisciplinary voice against global crises*. Palgrave.

Leyva, E. W. A., Beaman, A., & Davidson, P. M. (2017). Health impact of climate change in older people: An integrative review and implications for nursing. *Journal of Nursing Scholarship, 49*(6), 670–678. https://doi.org/10.1111/jnu.12346

Marquet, O., Hipp, J. A., & Miralles-Guasch, C. (2017). Neighborhood walkability and active ageing: A difference in differences assessment of active transportation over ten years. *Journal of Transport and Health, 7*(March), 190–201. https://doi.org/10.1016/j.jth.2017.09.006

Nesser, W., & Nesser, T. W. (2024). The impact of crises on older adults' health and function: An intergenerational perspective. In N. Asiamah, H. T. A. Khan, P. V. Dintrans, M. J. Koohsari, E. Mogaji, E. R. Vieira, R. Lowry, & H. K. Mensah (Eds.), *Sustainable neighbourhoods for ageing in place: An interdisciplinary voice against global crises*. Palgrave.

Öçal, M., & Kutlu, O. (2024). Two sides of the coin in aging in place: Neighbourhood safety and elder abuse. In N. Asiamah, H. T. A. Khan, P. V. Dintrans, M. J. Koohsari, E. Mogaji, E. R. Vieira, R. Lowry, & H. K. Mensah (Eds.), *Sustainable neighbourhoods for ageing in place: An interdisciplinary voice against global crises*. Palgrave.

Owen, N., Cerin, E., Leslie, E., duToit, L., Coffee, N., Frank, L. D., Bauman, A. E., Hugo, G., Saelens, B. E., & Sallis, J. F. (2007). Neighborhood walkability and the walking behavior of Australian adults. *American Journal of Preventive Medicine, 33*(5), 387–395. https://doi.org/10.1016/j.amepre.2007.07.025

Pani-Harreman, K. E., Bours, G. J. J. W., Zander, I., Kempen, G. I. J. M., & Van Duren, J. M. A. (2020). Definitions, key themes and aspects of "ageing in place": A scoping review. *Ageing and Society, 41*, 1–34. https://doi.org/10.1017/S0144686X20000094

Prempeh, C. (2024). A behavioural approach to sustainable neighbourhoods: A philosophical construction of a friendly neighbourhood. In N. Asiamah, H. T. A. Khan, P. V. Dintrans, M. J. Koohsari, E. Mogaji, E. R. Vieira, R. Lowry, & H. K. Mensah (Eds.), *Sustainable neighbourhoods for ageing in place: An interdisciplinary voice against global crises*. Palgrave.

Roy, B., Hajduk, A. M., Tsang, S., Geda, M., Riley, C., Krumholz, H. M., & Chaudhry, S. I. (2021). The association of neighborhood walkability with health outcomes in older adults after acute myocardial infarction: The SILVER-AMI study. *Preventive Medicine Reports, 23*, 101391. https://doi.org/10.1016/j.pmedr.2021.101391

Rudnicka, E., Napierała, P., Podfigurna, A., Męczekalski, B., Smolarczyk, R., & Grymowicz, M. (2020). The World Health Organization (WHO) approach to healthy ageing. *Maturitas, 139*(February), 6–11. https://doi.org/10.1016/j.maturitas.2020.05.018

Sadruddin, A. F. A., & Inhorn, M. C. (2020). Aging, vulnerability and questions of care in the time of COVID-19. *Anthropology Now, 12*(1), 17–23. https://doi.org/10.1080/19428200.2020.1760633

Sivam, A., Karuppannan, S., & Soltani, A. (2024). Ageing in place and built environment amenities at neighbourhood scale: The case of South Australia. In N. Asiamah, H. T. A. Khan, P. V. Dintrans, M. J. Koohsari, E. Mogaji, E. R. Vieira, R. Lowry, & H. K. Mensah (Eds.), *Sustainable neighbourhoods for ageing in place: An interdisciplinary voice against global crises*. Palgrave.

Strong, J., Varady, C., Chahda, N., Doocy, S., & Burnham, G. (2015). Health status and health needs of older refugees from Syria in Lebanon. *Conflict and Health, 9*(1), 1–10. https://doi.org/10.1186/s13031-014-0029-y

Sundquist, K., Eriksson, U., Kawakami, N., Skog, L., Ohlsson, H., & Arvidsson, D. (2011). Neighborhood walkability, physical activity, and walking behavior: The Swedish Neighborhood and Physical Activity (SNAP) study. *Social Science and Medicine, 72*(8), 1266–1273. https://doi.org/10.1016/j.socscimed.2011.03.004

Wahl, H. W., & Gerstorf, D. (2018). A conceptual framework for studying COntext Dynamics in Aging (CODA). *Developmental Review, 50*(August), 155–176. https://doi.org/10.1016/j.dr.2018.09.003

Wahl, H. W., & Gerstorf, D. (2020). Person-environment resources for aging well: Environmental docility and life space as conceptual pillars for future contextual gerontology. *Gerontologist, 60*(3), 368–375. https://doi.org/10.1093/geront/gnaa006

Yazdanpanahi, M., & Hussein, S. (2021). Sustainable ageing: Supporting healthy ageing and independence amongst older Turkish migrants in the UK. *Sustainability (Switzerland), 13*(18), 1–19. https://doi.org/10.3390/su131810387

Part I
Impacts of the Crises

Chapter 2
Ageing in Place: The Present and Future Social and Health Threats

Nestor Asiamah, Mohammad Javad Koohsari, and Ruth Lowry

Acronyms

ADLs	Activities of Daily Living
ATA	Activity Theory of Ageing
CODA	Context Dynamics in Ageing
COVID-19	Coronavirus Disease 2019
CTA	Continuity Theory of Ageing
DTA	Disengagement Theory of Ageing
IADLs	Instrumental Activities of Daily Living
NEWS	Neighbourhood Environment Walkability Scale
PANES	Physical Activity Neighbourhood Environment Scale
P-E	Person-Environment
SAN	Socially Active Neighbourhoods
SST	Socioemotional Selectivity Theory
SCT	Social Cognitive Theory

N. Asiamah (✉)
Division of Interdisciplinary Research and Practice, School of Health and Social Care, University of Essex, Colchester, Essex, UK

Department of Gerontology and Geriatrics, Africa Centre for Epidemiology, Accra, Ghana
e-mail: n.asiamah@essex.ac.uk; nestor.asiamah@ace-gh.org

M. J. Koohsari
Japan Advanced Institute of Science and Technology, Nomi, Ishikawa, Japan

Faculty of Sport Sciences, Waseda University, Tokorozawa, Saitama, Japan

School of Exercise and Nutrition Sciences, Deakin University, Burwood, VIC, Australia
e-mail: koohsari@jaist.ac.jp

R. Lowry
School of Sports, Rehabilitation, and Exercise Sciences, University of Essex, Colchester, Essex, UK
e-mail: r.lowry@essex.ac.uk

N. Asiamah et al. (eds.), *Sustainable Neighbourhoods for Ageing in Place*,
https://doi.org/10.1007/978-3-031-41594-4_2

2.1 Introduction

A superaged population has been defined as any population with at least 20% of older adults aged 65 years or higher (Muramatsu & Akiyama, 2011; World Health Organization, 2019). Current projections suggest that the world's population will be "superaged" by 2050. (McCurry, 2015; Muramatsu & Akiyama, 2011). Ageing is a natural phenomenon and can, therefore, not be avoided, especially in contemporary times when life expectancy continues to increase while fertility declines (Kpessa-Whyte, 2018; World Health Organization, 2019). Given the current trend of ageing, interventions aimed at enabling healthy ageing are imperative.

Healthy ageing is the process of developing and maintaining the functional capacity that enables optimal health or well-being in the old age (Rudnicka et al., 2020). This definition overlaps with Peel and colleagues' definition of successful ageing, which is high physical, psychological, and social functional ability in old age without major diseases (Peel et al., 2004). Both healthy and successful concepts concern a process in which physiological and cognitive changes accompanied by ageing are delayed or slowed down. A flagship concept often used closely with successful and healthy ageing is "ageing in place" (Asiamah et al., 2023; Pani-Harreman et al., 2021).

Researchers have expressed divergent views regarding what should be the ideal definition of ageing in place (Pani-Harreman et al., 2021), but gerontologists (Gan et al., 2020; Pani-Harreman et al., 2020) agree that ageing in place is basically about ageing in good health at home or maintaining a sense of attachment to one's neighbourhood. This framing of ageing in place connotes that one's home and neighbourhood, as opposed to residential care homes and related facilities, constitute the ideal place to age. Ageing in place can be made less realistic by prevailing "global crises", namely, climate change, radical industrialisation, the outbreak of infectious diseases, and systemic violence.

Several studies (Bains & Turnbull, 2019; Barchielli et al., 2022; Brockie & Miller, 2017; Quinn et al., 2022) have discussed how one or two of these crises can adversely affect ageing and the health of older adults, but none of them has considered the four crises at a time. This chapter, therefore, aims to formulate a theoretical framework as a lens for understanding the four crises as a force that can make ageing in place problematic, unrealistic, or even impossible. This chapter has seven sections, with the first section (i.e. the introduction) providing the rationale of the chapter. The second section recounts the four crises as they occur in a global context, whereas the third section provides a tripartite theoretical synthesis. The fourth section draws on the theoretical synthesis to propose a heuristic by which scholars and policymakers can understand the concept of "diminishing life space". The fifth section delineates implications for ageing in place, whereas the sixth and seventh sections present the limitations and conclusions, respectively.

2.2 An Overview of The Four Global Crises

Since the First World War, the world has been a more turbulent place characterised by natural and man-made disasters. Today, no country or region is free from disasters; even some of the most peaceful countries, including New Zealand, recently experienced a terrorist explosion in Canterbury (Battersby & Ball, 2019) and remain vulnerable to climate change events, including earthquakes and volcanic eruptions (Johnston et al., 2011). The Coronavirus 2019 disease (COVID-19) has been associated with mental health declines at the population level (Abbaspur-Behbahani et al., 2022; Kim & Jung, 2021) and left many people with prolonged grief and social trauma. This impact of COVID-19 has been coupled with macroeconomic difficulties and radical industrialisation fuelled by national urbanisation interests (Dimitriou et al., 2008; Li et al., 2018). If we were to look ahead through these circumstances, we would see little growth and peace but more turmoil and uncertainty.

We define radical industrialisation as the development of businesses, enterprises, or industries in a country on a large scale in a way that threatens nature (e.g. green space, animals, and rivers) embedded in neighbourhoods, livelihoods, and the normalcy of lives in communities. This type of industrialisation is a combination of legal and illegal economic activities undertaken by individuals, businesses, and even governments (Li et al., 2018). It has become commonplace in contemporary times due to increasing competition among individuals, businesses, and countries for success in a world where resources are diminishing due to population growth. Radical industrialisation can complement natural events, such as flooding, pollution, and hurricanes, to disturb the provision of essential services and destroy homes and public spaces necessary for a normal life (Barchielli et al., 2022). Uncontrolled industrial activities, which are the core of radical industrialisation (Barchielli et al., 2022; Dimitriou et al., 2008), can obliterate environmental greenery and other natural attributes of neighbourhoods (e.g. lakes and rivers) that support health-seeking behaviours. This adverse impact of radical industrialisation can worsen at the outbreak of an infectious disease.

In recent years, COVID-19 has affected the lives of most people globally and has garnered much attention; however, it is just one of many infectious diseases that have had wide ranging consequences. Ebola, monkeypox, and the flu are a few other infectious diseases that can be as dangerous as COVID-19. What makes these diseases a substantive threat to community life is their ability to break social bonds and turn socially vibrant communities into ghost towns due to their contagion. For example, COVID-19 made it necessary for residents to break away from their families and friends and abandon public spaces (Lewis et al., 2023; Neves et al., 2023). For a time, many neighbourhoods were not used, and social and economic groups (e.g. keep-fit clubs, churches, and businesses) became disintegrated, which resulted in social disengagement and mental health declines (Asiamah et al., 2021b; Kim & Jung, 2021; Lewis et al., 2023). Infectious diseases have compounded the burden of mortality and disease due to the increased vulnerability of people living with non-communicable diseases, such as cancer and cardiovascular diseases (Hu et al., 2023;

Mai et al., 2020; Oberoi et al., 2019). The economic, social, and health implications of the pandemic were enormous (Ross, 2020), so many governments were eager to lift restrictions, such as lockdowns and social distancing.

The society is expected to become more volatile as individuals struggle for financial security and countries crave recognition within geopolitics. Struggles among countries for world power have resulted in wars (Barchielli et al., 2022), whereas the interest of individuals (e.g. politicians) to enhance their financial security has heralded violence within communities (Benjaminsen et al., 2009; Issifu et al., 2022). The Russia-Ukraine conflict is a recent example of how conflict can displace families, causing untold economic hardships and breaking the fabric of Ukrainian communities (Barchielli et al., 2022; Oleksiyenko et al., 2021; Quinn et al., 2022). In many countries and regions (e.g. the Middle East), revolts among the youth and radical groups in response to the activities of federal governments have caused fear and panic (Bayat, 2013; Goodwin, 2011). Recurrent political unrest, especially during elections, around the world has become more prevalent (Goldsmith, 2015; Onapajo, 2014). As a result, residents in many countries and communities may see absolute peace as a mirage, more so in contexts periodically ravaged by climate change events.

Climate change is a big challenge for the world, though the urgency of its remedy is not accepted by all (Leiserowitz, 2005). Ocean acidification, hurricanes, flooding, landslides, drought, and extreme weather are major events associated with climate change. Ocean acidification threatens aquatic life and may significantly result in a lower supply of fish and seafood. Extreme weather disrupts air, sea, and land travel (Chen & Mahmassani, 2015; Hyland et al., 2018) and increases public health risks, such as skin cancer and asthma (Kovats & Kristie, 2006). Hurricanes, landslides, and flooding have rendered families and individuals homeless and caused the death of many (Benevolenza & DeRigne, 2019; Malik et al., 2018). Climate change is a global problem as it accompanies events (e.g. landslides, hurricanes, flooding, and heatwaves) that affect every country, though some countries are more impacted (Issifu et al., 2022; Leiserowitz, 2005). The universality of the radical industrialisation (Barchielli et al., 2022), war or violence (Lim et al., 2022), and infectious disease outbreaks (Lewis et al., 2023) has also been acknowledged. If so, it is imperative to know how these events affect communities and segments of the population most vulnerable to them. For theoretical reasons discussed in the next section, older adults are among the most vulnerable.

2.3 The Global Crises Threaten Healthy Ageing: What Does Theory Say?

Ageing is a natural phenomenon involving changes in human physiology; hence, some commentators recognise ageing as physiological changes characterised by a decline in functional ability over the life course (Asiamah et al., 2021a; Peel et al.,

2004). Sarcopenia (i.e. a loss of bone and muscle mass) and greying of the hair are some primary changes associated with the ageing process (Asiamah et al., 2021a; Sezgin et al., 2019).

2.3.1 Ageing, Physiological, and Cognitive Changes

One of the basic physiological changes accompanied by ageing that is recognised by the Disengagement Theory of Ageing (DTA) is a decline in functional capacity over time (Asiamah, 2017; Asiamah et al., 2023). This change is due to a gradual loss of bone and muscle mass across the lifespan, which results in frailty and limits physical and social activities. Frailty is the impairment and limitations in physical performance including Activities of Daily Living (ADLs) and Instrumental Activities of Daily Living (IADLs) (Dong et al., 2017; Sezgin et al., 2019). ADLs are self-care activities performed at home, such as bathing and toileting, whereas IADLs are social and physical activities performed beyond the individual's home, such as walking to a nearby supermarket (Sezgin et al., 2019). According to the DTA, the individual loses the ability to perform these activities (i.e. functional ability) during the ageing process due to the onset and progression of frailty.

Frailty is not only about physiological changes but also includes cognitive impairments, some of which may emanate from neurodegenerative disorders such as dementia (Asiamah et al., 2021a; Hoogendijk et al., 2018). This is to say that functional declines stemming from the ageing process include both physiological and cognitive impairments (Flaxman et al., 2017). The DTA asserts that these changes are unavoidable in the ageing process and are, therefore, bound to limit what older adults can physically do in later life. To illustrate, the ability of an individual to complete a marathon would dwindle and become less achievable in later life; this individual may be unable to sustain walking and related activities due to frailty.

The Continuity Theory of Ageing (CTA) and Activity Theory of Ageing (ATA) are closely related theories that are at odds with the DTA. These theories argue that though people lose functional ability in the ageing process, they can adapt past experiences to cope with changing life circumstances, including physiological declines. Learning through the ageing process and adapting past behaviours as well as experiences enable the individual to maintain their involvement in social and physical activities in later life. Thus, while the DTA asserts that the ability to perform physical tasks in older age is relatively low, the CTA and ATA posit that this ability can be maintained. Noteworthy is the fact that all these life course theories recognise a gradual decline in physical functional ability as an attribute of the ageing process. A review of these theories (Asiamah, 2017) has suggested that the imports of the DTA, as well as the ATA and CTA, are valid in different contexts. For instance, the maintenance of physical functional ability in old age is possible in settings where the physical and social environments (e.g. availability of walkable neighbourhoods and healthcare) are well designed to support health-seeking behaviours and optimal health. This idea highlights the role of the built environment in ageing.

2.3.2 The Role of the Built Environment in Ageing in Place

Since 1951 when the first person-environment (P-E) fit model was developed (Asiamah et al., 2023; Wahl & Gerstorf, 2018), the framing of the ideal place to age has gained remarkable momentum. Two landmark models developed by Powell M. Lawton and Marjorie H. Cantor (Cantor, 1975; Lawton & Simon, 1968) have been very influential in this regard; Lawton and Simon propounded the environment docility framework, whereas Cantor proposed the life-space concept, both of which recognise the role of the environment in health-seeking behaviours (e.g. social engagement and walking) and healthy ageing. The environment docility concept suggests that the less competent and resourceful people are, the greater the impact of environmental factors (e.g. global crises) on them.

The docility hypothesis implies that incompetence is implicit in poor physical health or a low functional ability and any factor that compels the individual to seek isolation away from neighbourhood resources, such as parks, services, and social networks (Wahl & Gerstorf, 2020). The life space concept, on the other hand, partly asserts that the use of contextual resources, such as services, depends on proximity to these resources (Asiamah et al., 2023; Wahl & Gerstorf, 2020). In later life, proximity may more strongly affect neighbourhood utilisation owing to physiological limitations accompanied by ageing. Both the docility and life space concepts agree that health and health-seeking behaviours are a function of whether the neighbourhood offers the right attributes (e.g. pavements, services, and social networks) and the ability of the individual to use these attributes. Physical performance and resources (e.g. high income) are part of the individual's capability to use these neighbourhood resources to maintain health-seeking behaviours and optimal health.

The docility hypothesis recognises that older adults are more likely to lack the ability (e.g. functional capacity) and resources (e.g. social networks and income) to use neighbourhood resources (Wahl & Gerstorf, 2018, 2020). In other words, older adults are less capable (e.g. less competent) of meeting environmental demands, which concern the availability of the foregoing attributes. In the late 1980s, Powell M. Lawton built on the environmental docility concept with the proactivity hypothesis (Wahl & Gerstorf, 2020), which states that ageing individuals can shape their environment to facilitate resource availability and use. This argument resonates with the import of the ATA and CTA that ageing people can adapt past experiences and behaviours to maintain engagement with life in the community. Adaptation of past experiences is a route to shaping one's social and physical environment by maintaining neighbourhood-level resources, such as social networks and support (Asiamah et al., 2021a).

The Context Dynamics in Ageing (CODA) framework integrates the above P-E fit models and emphasises the role of psychosocial factors, such as safety and peace in ageing (Wahl & Gerstorf, 2018). This model adds that psychosocial factors are an essential part of neighbourhoods supporting health and health-seeking behaviours over the life course. This reasoning was more recently elaborated and mirrored in another integrated model, the Socially Active Neighbourhood (SAN) framework

(Asiamah et al., 2023). The SAN contends that the maintenance of optimal health in the ageing process requires the ability to use sociable neighbourhoods comprising psychosocial factors (e.g. peace, safety, and social cohesion) and walkability factors (e.g. high residential density, mixed land use, and street connectivity). The SAN as a model views healthy ageing as a function of psychosocial factors, walkability, and personal factors (e.g. functional ability and income). This framing of an age-friendly context conveys why stakeholders should be worried about global crises. As the next section explains, older adults may not have enough resources to maintain resilience and cope in the face of global crises.

2.3.3 The Theoretical Basis of Coping and Resilience Against the Crises

Apart from the CTA and ATA, some other theories recognise the ability of older adults and ageing people to learn and adapt in response to ageing. An example is the Social Cognitive Theory (SCT) (Kosteli et al., 2016), which avers that individuals, through the ageing process observe models (other people) and replicate behaviours performed by these models, depending on how beneficial these behaviours are perceived to be. This point of view overlaps with the ATA and CTA, both of which recognise adaptive behaviour as a requirement for successful ageing. It implies that a child who grew up with parents who regularly exercised is likely to maintain physical activity in later life. As implied by the Socioemotional Selectivity Theory (SST) (Isaacowitz et al., 2003), nevertheless, every human ability can be self-limited or constrained.

The SST echoes the argument of the DTA by asserting that ageing is associated with a change in life goals which impels older adults to focus on activities and choices that matter to them (Cleveland & Agbeke, 2019; Löckenhoff & Carstensen, 2004). This change in life goals may be due to older adults' *future time perspective*, which is about their recognition of how short their remaining life is (Cleveland & Agbeke, 2019). Given this recognition, older adults are selective of their social activities and engage only in events perceived as rewarding. They aim to keep social networks in their age group and avoid social activities with individuals outside their age group. Older adults, therefore, limit their daily activities to a few goals tied to their future time perspective. This self-limiting attitude is a major barrier to adaptive behaviour and can be expected to weaken resilience and coping amidst global crises.

The discussion so far shows two divergent schools of thought: the idea that ageing is associated with a decline in individual resources and abilities, on the one hand, and the belief that people can maintain these abilities and resources in old age, on the other hand. As mentioned earlier, both lines of reasoning are plausible and valid, depending on the context involved. Even so, both schools of thought agree that the ageing individual needs to adapt previous experiences and behaviours to overcome the challenges accompanied by ageing, including frailty. This stark convergence of the DTA and ATA suggests that ageing people need the ability to adapt and resist the natural challenges accompanied by ageing. We argue that this ability to adapt previous behaviours and resist evolving challenges requires resilience (Huisman et al., 2017), which we define as the ability to quickly overcome or recover from difficulties.

Since ageing is inevitable, its challenges, such as frailty and a decline in functional capacity, are unavoidable difficulties that individuals must adjust. Yet, in contexts where one or more of the global crises are experienced, the ability of individuals to adapt past behaviours and experiences to cope with age-related difficulties would be low. Suffice it to say that global crises can compound the biological and environmental factors that facilitate a decline in health and physical functional ability over the life course. These crises can also result in disability or health problems (e.g. injuries from war or violence) that can catalyse or worsen age-related morbidities. Diseases from extreme weather (e.g. heatwaves), injuries from war, and the join impact of infectious and noncommunicable diseases, such as cancer, can push older adults to the limit of their coping ability (Barchielli et al., 2022; Gamble et al., 2013), suggesting that older adults' resilience may not be enough against the enormity of events from one or more of the global crises. The consequence of this situation is what we call a "diminishing life space".

2.4 The Global Crises and Diminishing Life Space

2.4.1 Diminishing Life Space Amidst the Global Crises

Drawing on the argument of the DTA, we reason that the individual would withdraw gradually from society in the way of social isolation due to a change in life goals and the decline in physical functional ability as well as personal resources such as social connections and income. The SST implies that ageing people, based on their future time perspective, would avoid many social activities to focus on only choices that matter (e.g. spending more time with grandchildren), thereby limiting their use of neighbourhood resources. Intentional and purposeful withdrawal from society is behaviour necessitated by the ageing process and limits access to "life space", which we define as the neighbourhood (i.e. the distance between 10 and 15 min walk away from home) (Asiamah et al., 2023; Sallis et al., 2010) characterised by services, streets, parks, and social ties (e.g. friends). A diminishing life space is a gradual decline in the individual's access to the neighbourhood and its pro-health attributes (e.g. services, parks, and social ties) due to social isolation or disuse of neighbourhood resources.

The above bioecological models also recognise diminishing life space analogous to social isolation attributable to a loss of functional capacity in the ageing process. The proactivity hypothesis, the CTA, and ATA premise in opposition to this reasoning that a reduction in the size of life space can be delayed or avoided. Nonetheless, whether the individual can delay or avoid a reduction in life space in the ageing process depends on several factors (e.g. the physical environment, physical health status, and individual or household income) and the interaction between them. Of interest is the physical environment, which influences whether people would use neighbourhood resources through social and physical activities (Wahl & Gerstorf, 2018, 2020). Because the physical environment is subject to global crises, the prevalence of any of these crises in any context would encourage or necessitate social isolation and, consequently a smaller life space. For instance, violence and heatwaves due to climate change may compel older adults to remain in isolation even if they have the resources needed to maintain engagement with life (Hachem et al., 2022; Oleksiyenko et al., 2021). Air and noise population from industrial activities may also discourage social engagement and the utilisation of contextual resources. Hence, ageing adults need adaptive behaviour to overcome the physiological, social, and economic challenges accompanied by ageing and optimum resilience against the global crises. Suffice it to say that resilience must be optimised at a high intensity of crises. In a situation where older adults do not have enough resilience and coping ability against crises, possibly due to age-related factors (e.g. a lack of social support and frailty), they may be compelled to move out of their preferred homes to unfamiliar places.

Each of the global crises can obliterate the key factors (e.g. streets, parks, services, safety, and peace) that make life space inhabitable, but their joint impact on community life can be expected to be more alarming. Recent commentaries (Asiamah, 2022; Barchielli et al., 2022; Quinn et al., 2022) have unfolded how global crises can interact to ruin even the most peaceful and socially stable communities, forcing older adults and other vulnerable groups to seek isolation or refuge in places far away from their preferred home. Other commentators (Asiamah, 2022; Barchielli et al., 2022) are also concerned about the likelihood of neighbourhoods affected by two or more of the crises losing their sustainability. We define neighbourhood sustainability as the preservability of neighbourhood attributes (e.g. essential services), including the natural ones (e.g. parks, gardens, forests, and animals) that are necessary for a normal life for all segments of the population. The sustainability of neighbourhoods can be expected to significantly reduce in the aftermath of a war, hurricane, landslide, unauthorised industrial activity (e.g. destroying a park to build a factory), or a combination of these. The occurrence of one or more of the crises at a time can rapidly change the social status of communities over time.

2.4.2 *The Global Crises and the "Social Status" of Neighbourhoods*

In this chapter, social status encompasses the desired situation of peace, economic agility, and congeniality that everyone desires in a community. This status is a recipe for economic growth or fosters a social and economic climate where the values of society are co-created. We classify any community or neighbourhood with this status as a "haven". Ideally, every country and community should be a haven to support health-seeking behaviours, inclusion, economic development, and ageing in place. As climate change intensifies and many neighbourhoods become vulnerable to it and other global crises, nevertheless, the sustainability of havens is less likely. While this may sound pessimistic, it reinforces the importance of safeguarding haven-like life spaces.

Havens may lose their desirable status and plunge into distress caused by any of the four crises, which is why we captured an "unstable haven" in Table 2.1. An unstable haven is a context where at least one of the crises occurs several times a year, but residents enjoy some attributes of a haven intermittently. Because social situations are unstable in such places, their residents are generally anxious and constantly sit on tenterhooks. Some of its residents may leave for safer places if they can, but some may remain, hoping for a permanent recovery from the crisis. Unstable havens can experience high economic losses during episodes of the crises and are, therefore, likely to degenerate into an infernal, the worse status a context can have.

An infernal is a context where one or more of the crises occur frequently, and residents do not enjoy any moment of respite from the crises. The crises may occur together, or more than one may constantly occur, resulting in a depletion of economic and neighbourhood resources that encourage health-seeking behaviours. The age-related health problems of older adults may be compounded by the crises, which signifies the vulnerability of people in an infernal to disability and unnatural death. At the peak of a pandemic, war, or natural disaster, residents with the financial capability or support from stakeholders relocate or seek refuge elsewhere (Bazaluk & Balinchenko, 2020; Strong et al., 2015), which connotes that residents in an infernal would leave for safer places if they have the means or support to do so. Thus, ageing in one's neighbourhood or home is most unlikely in an infernal, though interventions rolled out by governments and civil society can revert it to a better status, possibly in what we call "a continuum of diminishing life space".

2.4.3 A Continuum of Diminishing Life Space and Ageing in Place

A diminishing life space can be a continuum whereby havens gradually deteriorate into an infernal. We illustrate this continuum with a heuristic shown in Fig. 2.1. As the figure shows, the continuum may start with a haven where there is optimum freedom from the global crises. Ultimately, every country and neighbourhood should strive to be at this stage of the cycle and routinely implement strategies to retain this social status. When one or more of the crises occur, local authorities may be overwhelmed by economic loss and peace deprivation. They may, therefore, be unable to immediately recover from the crises. Depending on the intensity of the global crises, the neighbourhood or context may not fully recover before another

Table 2.1 Attributes of three types of human ecosystems implied by the prevalence of the crises

Type	General condition	Psychosocial attributes	Socioeconomic attributes	Physical attributes	Impact on health and health-seeking behaviours	Space retention and implications for ageing in place
Haven	Absence or rare occurrence of any of global crises and interventions are put in place to mitigate their impact even if they occasionally occur	Trust, a sense of community, and social cohesion	Economic stability, high employment rate, and availability of financial support for more vulnerable groups during crises	Neighbourhoods retain their natural features (e.g. greenery), are not made dirty or uninhabitable by the crises (e.g. floods), and support social and physical activities	Older adults can easily access services (e.g. healthcare) and engage in physical and social activities to maintain health. Residents can maintain optimal health by using services and other attributes	Older adults will always live in the neighbourhood if they want, and governmental policies may support them to remain at home. Older adults are unlikely to relocate or die due to the crises
Unstable haven[a]	This context alternatives between the conditions of a haven and an infernal	Anxiety and uncertainty	Economic instability, unstable employment rate, and unstable as well as limited support for vulnerable groups during the crises	Some physical attributes are available, but these are destroyed occasionally by the crises (e.g. floods, violence). Attributes replaced by stakeholders (e.g. government) are destroyed again later by the crises	Older adults are generally anxious and uncertain about the future, and they are unable to maintain health-seeking behaviours, such as social and physical activities due to the crises	Older adults are likely to leave the neighbourhood to seek refuge elsewhere, and they may not have the financial means to relocate. The extreme outcome is likely death attributable to the long-term impacts of the crises on mental and physical health

(continued)

Table 2.1 (continued)

Type	General condition	Psychosocial attributes	Socioeconomic attributes	Physical attributes	Impact on health and health-seeking behaviours	Space retention and implications for ageing in place
Infernal	One or more of the crises always occur (i.e. daily, weekly, or monthly)	Extreme vulnerability, panic, fear, and volatility	Economic deprivation, non-availability of support for vulnerable groups, and low employment rate, possibly due to the crises	Neighbourhoods lack natural features (e.g. greenery), services are unavailable or inaccessible, and sanitation is low due to one or more of the crises (e.g. flooding, vandalism, or violence)	Older adults lack support, remain in isolation, and may not be able to maintain health-seeking behaviours and optimal health; even if they don't relocate to places perceived to be safer, their quality of life may be low	Older adults would want to leave for safer places, but they may not have the economic means or may be trapped. The extreme outcome is disability or death from any of the crises (e.g. being killed during a war)

[a]This life space becomes a semi-haven if it deteriorates to an infernal or alternates between a haven and an infernal

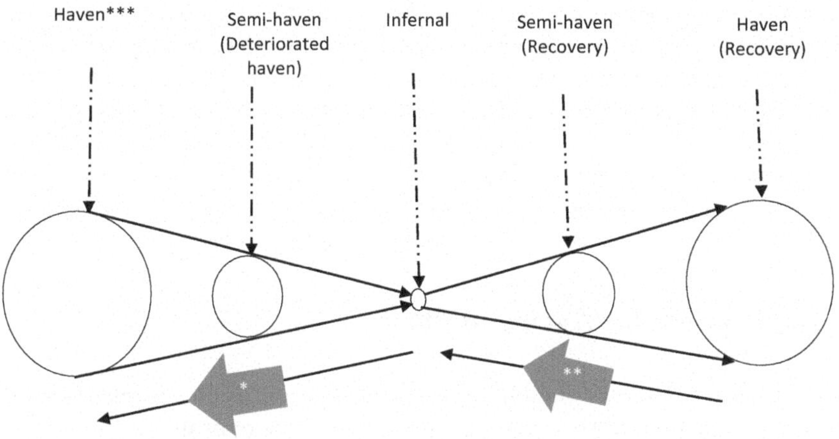

Note: This framework is the authors' construct; ***the continuum starts at a "haven"; **indicates a second decline in social status down to "infernal"; *indicates reversed (second) recovery upward from an "infernal"; a recovery is ideally co-created by different stakeholders such as residents, governments, and civil society, and is a gradual process of restoring spaces affected by any of the crises to a haven.

Fig. 2.1 A continuum of diminishing space due to the crises and recoveries driven by interventions

unexpected event occurs. This development can change the status of a haven to a semi-haven generally characterised by attributes of an unstable haven. A semi-haven is the same as an unstable haven found in the continuum or cycle in the sense that it faces the risk of plunging further into an infernal and possibly recovering back to a better social status later. This ecosystem is likely to go through this cycle because of economic difficulties experienced during the crises (Ross, 2020) or a lack of sustainable interventions intended to prevent the crises or mitigate their impacts at the population level.

Attaining the status of an infernal through a semi-haven is reversible through interventions aimed at a recovery, which we define as a context's return to a better social status. As the second part of Fig. 2.1 shows, a recovery from an infernal through a semi-haven to a haven is possible, especially with interventions co-designed by local and international groups. As opined by some commentators (Klenk et al., 2015), support from international organisations and groups (e.g. United Nations and high-income countries) is necessary to economically reinvigorate any context that has experienced a crisis or a decline in social status over time. The arrow with a single asterisk in the heuristic indicates a recovery from the first cycle of social status decline, whereas the other arrow with a double asterisk illustrates a backward recovery after the context had reverted to an infernal after completing the first cycle of recovery. Thus, Fig. 2.1 depicts the possibility of an ecosystem repeating the cycle if interventions are relaxed or withdrawn while one or more of the crises remain.

Instability in social status can be characterised by a movement through the cycle or between two social statuses. An unstable haven, though, alternates between only a haven and a semi-haven, and this type of confined instability is typical of contexts alternating between contrasting regimes, a government fostering freedom from the crises through interventions (e.g. peace treaties) (Bell & O'Rourke, 2007; Khan, 2021), and a regime relaxing interventions or giving way to the escalation of violence or any of the crises (Khan, 2021). Deductively, interventions must be sustained if countries want to maintain their desired social status or avoid going down the cycle, since the sustainability of the ideal life space has implications for ageing in place.

2.5 Implications for Ageing in Place

The consequence of a diminishing life space, which occurs when a neighbourhood moves down from a haven to an infernal due to one or more of the crises, is residents' "psychological distance". We define psychological distance as residents' thoughts and feelings about remaining indoors or leaving their homes for places less exposed to the crises. It is about residents affected by the crises contemplating a relocation for refuge or safety. Psychological distance increases as people become more anxious and inclined to leave their current neighbourhoods. An increase in the individual's psychological distance is more probable in contexts experiencing multiple crises. Recognising their cognitive and physiological limitations, older adults would experience a longer psychological distance compared to younger people. Thus, the length of a psychological distance can be explained by the intensity of the crisis, the number of crises occurring at a time, and the vulnerabilities being experienced by residents.

Two types of psychological distance may affect ageing in place. The first of these is neighbourhood psychological distance, which is limited to the individual's immediate neighbourhood and occurs when residents think about taking cover at home in the way of social isolation to avoid being harmed by the crises. This type of psychological distance was experienced by residents who decided to socially isolate at the peak of the COVID-19 pandemic, though there were no lockdowns in place (Asiamah et al., 2021c; Kim & Jung, 2021). The second type of psychological distance is out-of-home psychological distance, which occurs when residents contemplate relocating to another neighbourhood or context for safety or refuge. With this type of psychological distance, residents think about moving out of their homes to potentially less homely nearby or distant places that are not as exposed to the crises as their neighbourhood. This type of psychological distance was experienced by residents who may have travelled out of their countries or neighbourhoods to places less affected by COVID-19 (Ullah et al., 2021). Residents, especially those with underlying health conditions (e.g. asthma), may be compelled by atmospheric dust and noise from factory activities to relocate. Out-of-home psychological distance is longer than neighbourhood psychological distance, because it can lead to relocation to places beyond the individual's home and neighbourhood.

Though the psychological distance is only a feeling or thought about socially isolating or leaving one's home, it can result in the self-evacuation of residents with the physical and financial means to travel. It, thus, represents a mental state that threatens homeliness and people's attachment to their place of residence. From this perspective, psychological distance evoked or worsened by one or more of the crises can undermine the two fundamental life conditions necessary for people to remain in their neighbourhoods: (1) a sense of attachment to place (Pani-Harreman et al., 2020) and (2) stability and enjoyability of life in the individual's home, which is generally better than any other unfamiliar place (Fischl et al., 2020). We capture the implications of this situation in the following postulates:

Postulate 1 – Life space may diminish as the individual withdraws from society in the way of social disengagement in the ageing process. The individual may lose access to the social and physical environment due to social disengagement necessitated by a change in life goals (e.g. deciding to spend more time with loved ones indoors) or a decline in physical and cognitive abilities. In later life, the global crises are more likely to increase social isolation (or disuse of life space) by complementing older adults' physiological limitations (e.g. frailty and low functional ability).

Postulate 2 – Life space can further diminish as a neighbourhood goes down the heuristic (cycle, see Fig. 2.1) amidst the crises, causing a psychological distance and a possible relocation of the individual to other potentially less homely contexts that seem safer. Even if people do not leave their homes, their fear of the unexpected and psychological distance can terminate their sense of attachment and enjoyment of life. Residents, in essence, may psychologically or emotionally age out of their neighbourhoods and homes.

Postulate 3 – The occurrence of a crisis is enough to cause psychological distance and subsequently a loss of sense of attachment to one's home and neighbourhood. A loss of a sense of attachment is more likely with the number and intensity of crises occurring and may lead to relocation of individuals. The crises may contribute to poor health, frailty, or disability, which may call for older residents being taken into nursing homes.

Postulate 4 – Neighbourhood sustainability is needed to avoid psychological distance and maintain a sense of attachment and health-seeking behaviours within life space. A neighbourhood's sustainability may fall as the number and intensity of the crises increase. Stakeholder interventions can maintain neighbourhood sustainability during or after the crises.

Postulate 5 – Optimum resilience is needed by individuals, especially older adults, to cope with crises. Still, the ability to cope with high resilience does not necessarily signify the absence of psychological distance. Due to their physiological and cognitive limitations, older adults may not exercise enough resilience to cope with global crises. Training, though, may improve the ability of people to build and maintain resilience against global crises (Musich et al., 2022). Personal factors, such as physical health status, income, and availability of social support, may determine a person's psychological distance and whether it will result in relocation or a loss of a sense of attachment.

Postulate 6 – A life space diminishes and becomes unstable or plunges down the heuristic if interventions are not rolled out and maintained to prevent the reoccurrence of the crises or to shield residents from their impact. Psychological distance may increase as functional and cognitive abilities decline in the ageing process, especially amidst global crises. Ideally, governments and other stakeholders must collaboratively work to prevent the occurrence of global crises or shield residents from their impacts. This effort must be maintained to prevent a reoccurrence or an escalation of the crises.

2.6 Limitations and Policy Implications

Our heuristic is not based on empirical data and is only used to describe potential changes a life space may undergo when experiencing one or more global crises. Yet, it can be used to determine a checklist of criteria (e.g. social and physical environmental characteristics shown in Table 2.1) that contexts should meet to be called a haven. This checklist may be used to locate a neighbourhood or wider context on the heuristic, possibly as part of an evaluation of the impact of interventions being implemented by stakeholders. Whether an intervention has improved the social status of a context over a defined period can be assessed and known by the checklist. This checklist, though, may be developed with a standard psychometric test (Streiner & Kottner, 2014), which involves the development of the relevant questions or items and an assessment of their psychometric properties (e.g. internal consistency, interclass correlation, test-retest reliability, discriminant validity, convergent validity, criterion-related validity, and face validity).

Due to their physiological limitations, older adults would feel more vulnerable to the global crises; hence, their psychological distance would be longer in every context. For this reason, governments and other stakeholders may prioritise the safety of older adults in future interventions, including planning cities resilient against global crises. The aim of stakeholders may be to roll out interventions (e.g. city design) that would make older adults feel relatively safe in their homes during crises.

Our framework only describes the three types of life spaces that may be the consequence of the crises. These types of life spaces and the heuristic describe a gradient of social statuses determined by the crises, which suggests that a haven is not necessarily completely free from the crises or related issues and an infernal is not always a hotbed of the crises. We could not mention specific examples (i.e. neighbourhoods and countries) of the three life spaces, but future research is encouraged to do this and to evidence the heuristic's practicality.

2.7 Conclusion

The four crises may affect human lives for a long time. Physiological and cognitive changes and a change in life goals necessitated by the ageing process compel people to limit their engagement with society, resulting in a smaller life space in later life. As the intensity of the crises increases in a neighbourhood or wider context, life space in later life would further shrink, making ageing in place less possible or even unrealistic. By their facilitation of a psychological distance, the crises can result in anxiety, fear, and economic difficulties that compel people, especially older adults, to lose a sense of attachment. Hence, even if people do not leave their homes for potentially less homely places believed to be safer, they may not age well in their homes. There is a need for governments, civil society organisations, and other stakeholders to co-develop sustainable interventions to the crises, enabling countries to avoid alternating between episodes of the crises or even going through the social status cycle.

Review Questions

1. Discuss at least two ways in which each of the four global crises may affect the lives of older adults?
2. List two major components of the neighbourhood that influence ageing in place. How do you think these components affect ageing in place? What is the relationship between the two components?
3. Describe the concept of "diminishing life space" and explain how it is influenced by the global crises.
4. Use a systems-thinking model to explain the social status of a neighbourhood and how it changes due to the global crises.
5. What is psychological distance, and which of the postulates describes it?

References

Abbaspur-Behbahani, S., Monaghesh, E., Hajizadeh, A., & Fehresti, S. (2022). Application of mobile health to support the elderly during the COVID-19 outbreak: A systematic review. *Health Policy and Technology, 11*(1). Elsevier B.V. https://doi.org/10.1016/j.hlpt.2022.100595

Asiamah, N. (2017). Social engagement and physical activity: Commentary on why the activity and disengagement theories of ageing may both be valid. *Cogent Medicine, 4*(1), 1289664. https://doi.org/10.1080/2331205x.2017.1289664

Asiamah, N. (2022, March). Walkable urban neighborhoods: The adverse effects of industrialization and climate change in developing countries. In *Sustainable urbanism in developing countries* (pp. 247–262). https://doi.org/10.1201/9781003131922-16

Asiamah, N., Conduah, A. K., & Eduafo, R. (2021a). Social network moderators of the association between Ghanaian older adults' neighbourhood walkability and social activity. *Health Promotion International*, 1–11. https://doi.org/10.1093/heapro/daaa156

Asiamah, N., Opuni, F. F., Mends-Brew, E., Mensah, S. W., Mensah, H. K., & Quansah, F. (2021b). Short-term changes in behaviors resulting from COVID-19-related social isolation and their influences on mental health in Ghana. *Community Mental Health Journal, 57*(1), 79–92. https://doi.org/10.1007/s10597-020-00722-4

Asiamah, N., Vieira, E. R., Kouveliotis, K., Gasana, J., Awuviry-Newton, K., & Eduafo, R. (2021c). Associations between older African academics' physical activity, walkability and mental health: A social distancing perspective. *Health Promotion International, September 2021.* https://doi.org/10.1093/heapro/daab093

Asiamah, N., Bateman, A., Hjorth, P., Khan, H. T. A., & Danquah, E. (2023). *Socially active neighborhoods: Construct operationalization for aging in place, health promotion and psychometric testing* (pp. 1–10).

Bains, K. K., & Turnbull, T. (2019). Improving health outcomes and serving wider society: The potential role of understanding and cultivating prosocial purpose within Health Psychology Research and Practice to address Climate Change and Social Isolation and Loneliness. *Frontiers in Psychology, 10*(JULY), 1–11. https://doi.org/10.3389/fpsyg.2019.01787

Barchielli, B., Cricenti, C., Gallè, F., Sabella, E. A., Liguori, F., Da Molin, G., Liguori, G., Orsi, G. B., Giannini, A. M., Ferracuti, S., & Napoli, C. (2022). Climate changes, natural resources depletion, COVID-19 pandemic, and Russian-Ukrainian War: What Is the impact on habits change and mental health? *International Journal of Environmental Research and Public Health, 19*(19). https://doi.org/10.3390/ijerph191911929

Battersby, J., & Ball, R. (2019). Christchurch in the context of New Zealand terrorism and right wing extremism. *Journal of Policing, Intelligence and Counter Terrorism, 14*(3), 191–207. https://doi.org/10.1080/18335330.2019.1662077

Bayat, A. (2013). The Arab spring and its surprises. *Development and Change, 44*(3), 587–601. https://doi.org/10.1111/dech.12030

Bazaluk, O., & Balinchenko, S. (2020). Dynamic coordination of internal displacement: Return and integration cases in Ukraine and Georgia. *Sustainability (Switzerland), 12*(10). https://doi.org/10.3390/su12104123

Bell, C., & O'Rourke, C. (2007). The people's peace? Peace agreements, civil society, and participatory democracy. *International Political Science Review, 28*(3). https://doi.org/10.1177/0192512107077094

Benevolenza, M. A., & DeRigne, L. A. (2019). The impact of climate change and natural disasters on vulnerable populations: A systematic review of literature. *Journal of Human Behavior in the Social Environment, 29*(2), 266–281. https://doi.org/10.1080/10911359.2018.1527739

Benjaminsen, T. A., Maganga, F. P., & Abdallah, J. M. (2009). The Kilosa killings: Political ecology of a farmer-herder conflict in Tanzania. *Development and Change, 40*(3), 423–445. https://doi.org/10.1111/j.1467-7660.2009.01558.x

Brockie, L., & Miller, E. (2017). Older adults' disaster lifecycle experience of the 2011 and 2013 Queensland floods. *International Journal of Disaster Risk Reduction, 22*(August 2016), 211–218. https://doi.org/10.1016/j.ijdrr.2016.08.001

Cantor, M. H. (1975). Life space and the social support system of the inner city elderly of new york. *Gerontologist, 15*(1), 23–27. https://doi.org/10.1093/geront/15.1_Part_1.23

Chen, R. B., & Mahmassani, H. S. (2015). Let it rain: Weather effects on activity stress and scheduling behavior. *Travel Behaviour and Society, 2*(1), 55–64. https://doi.org/10.1016/j.tbs.2014.08.001

Cleveland, J. N., & Agbeke, D. V. (2019). *Socioemotional selectivity theory lifespan perspectives on job performance, performance appraisal/Management and creative performance.*

Dimitriou, E., Karaouzas, I., Sarantakos, K., Zacharias, I., Bogdanos, K., & Diapoulis, A. (2008). Groundwater risk assessment at a heavily industrialised catchment and the associated impacts on a peri-urban wetland. *Journal of Environmental Management, 88*(3), 526–538. https://doi.org/10.1016/j.jenvman.2007.03.019

Dong, L., Liu, N., Tian, X., Qiao, X., Gobbens, R. J. J., Kane, R. L., & Wang, C. (2017). Reliability and validity of the Tilburg Frailty Indicator (TFI) among Chinese community-dwelling older people. *Archives of Gerontology and Geriatrics, 73*(July), 21–28. https://doi.org/10.1016/j.archger.2017.07.001

Fischl, C., Lindelöf, N., Lindgren, H., & Nilsson, I. (2020). Older adults' perceptions of contexts surrounding their social participation in a digitalized society – An exploration in rural communities in Northern Sweden. *European Journal of Ageing, 17*(3), 281–290. https://doi.org/10.1007/s10433-020-00558-7

Flaxman, S. R., Bourne, R. R. A., Resnikoff, S., Ackland, P., Braithwaite, T., Cicinelli, M. V., Das, A., Jonas, J. B., Keeffe, J., Kempen, J., Leasher, J., Limburg, H., Naidoo, K., Pesudovs, K., Silvester, A., Stevens, G. A., Tahhan, N., Wong, T., Taylor, H., et al. (2017). Global causes of blindness and distance vision impairment 1990–2020: A systematic review and meta-analysis. *The Lancet Global Health, 5*(12), e1221–e1234. https://doi.org/10.1016/S2214-109X(17)30393-5

Gamble, J. L., Hurley, B. J., Schultz, P. A., Jaglom, W. S., Krishnan, N., & Harris, M. (2013). Climate change and older Americans: State of the science. *Environmental Health Perspectives, 121*(1), 15–22. https://doi.org/10.1289/ehp.1205223

Gan, D. R. Y., Fung, J. C., & Cho, I. S. (2020). Neighborhood experiences of people over age 50: Factor structure and validity of a scale. *Gerontologist, 60*(8), E559–E571. https://doi.org/10.1093/geront/gnz111

Goldsmith, A. A. (2015). Elections and civil violence in new multiparty regimes: Evidence from Africa. *Journal of Peace Research, 52*(5), 607–621. https://doi.org/10.1177/0022343315576014

Goodwin, J. (2011). Why we were surprised (Again) by the Arab spring. *Swiss Political Science Review, 17*(4), 452–456. https://doi.org/10.1111/j.1662-6370.2011.02045.x

Hachem, S., Ali, S., Al-Omari, S., Abi Chahine, M., Fahme, S. A., & Mehio Sibai, A. (2022). "Older people tend to be invisible": A qualitative study exploring the needs and inclusion of older Syrian refugees in the context of compounding crises in host country, Lebanon. *Conflict and Health, 16*(1), 1–15. https://doi.org/10.1186/s13031-022-00496-4

Hoogendijk, E. O., Heymans, M. W., Deeg, D. J. H., & Huisman, M. (2018). Socioeconomic inequalities in frailty among older adults: Results from a 10-Year longitudinal study in the Netherlands. *Gerontology, 64*(2), 157–164. https://doi.org/10.1159/000481943

Hu, W., Zhai, C., Sun, H., Gong, X., Cui, L., Cai, L., Zong, Q., Yu, G., Wang, F., & Zou, Y. (2023). The global burden of disease attributable to metabolic risks in 204 countries and territories from 1990 to 2019. *Diabetes Research and Clinical Practice, 196*(October 2022), 110260. https://doi.org/10.1016/j.diabres.2023.110260

Huisman, M., Klokgieters, S. S., & Beekman, A. T. F. (2017). Successful ageing, depression and resilience research; A call for a priori approaches to investigations of resilience. *Epidemiology and Psychiatric Sciences, 26*(6), 574–578. https://doi.org/10.1017/S2045796017000348

Hyland, M., Frei, C., Frei, A., & Mahmassani, H. S. (2018). Riders on the storm: Exploring weather and seasonality effects on commute mode choice in Chicago. *Travel Behaviour and Society, 13*(April 2017), 44–60. https://doi.org/10.1016/j.tbs.2018.05.001

Isaacowitz, D. M., Smith, T. B., & Carstensen, L. L. (2003). Socioemotional selectivity and mental health among trauma survivors in old age. *Ageing International, 28*(2), 181–199. https://doi.org/10.1007/s12126-003-1023-7

Issifu, A. K., Darko, F. D., & Paalo, S. A. (2022). Climate change, migration and farmer–herder conflict in Ghana. *Conflict Resolution Quarterly, 39*(4), 421–439. https://doi.org/10.1002/crq.21346

Johnston, D., Tarrant, R., Tipler, K., Coomer, M., Pedersen, S., & Garside, R. (2011). Preparing schools for future earthquakes in New Zealand: Lessons from an evaluation of a Wellington school exercise. *Australian Journal of Emergency Management, 26*(1), 24–30.

Khan, A. (2021). Steps-to-war theory and interstate wars in the Middle East: Is state-sponsored terrorism another escalating step? *Journal of Asian and African Studies, 56*(7), 1521–1537. https://doi.org/10.1177/0021909620970573

Kim, H. H. S., & Jung, J. H. (2021). Social isolation and psychological distress during the COVID-19 pandemic: A cross-national analysis. *Gerontologist, 61*(1), 103–113. https://doi.org/10.1093/geront/gnaa168

Klenk, N. L., Meehan, K., Pinel, S. L., Mendez, F., Lima, P. T., & Kammen, D. M. (2015). Stakeholders in climate science: Beyond lip service? *Science, 350*(6262), 743–744. https://doi.org/10.1126/science.aab1495

Kosteli, M. C., Williams, S. E., & Cumming, J. (2016). Investigating the psychosocial determinants of physical activity in older adults: A qualitative approach. *Psychology and Health, 31*(6), 730–749. https://doi.org/10.1080/08870446.2016.1143943

Kovats, R. S., & Kristie, L. E. (2006). Heatwaves and public health in Europe. *European Journal of Public Health, 16*(6), 592–599. https://doi.org/10.1093/eurpub/ckl049

Kpessa-Whyte, M. (2018). Aging and demographic transition in Ghana: State of the elderly and emerging issues. *Gerontologist, 58*(3), 403–408. https://doi.org/10.1093/geront/gnx205

Lawton, M. P., & Simon, B. (1968). The ecology of social relationships in housing for the elderly. *The Gerontologist, 8*(2), 108–115. https://doi.org/10.1093/geront/8.2.108

Leiserowitz, A. A. (2005). American risk perceptions: Is climate change dangerous? *Risk Analysis, 25*(6), 1433–1442. https://doi.org/10.1111/j.1540-6261.2005.00690.x

Lewis, C., Phillipson, C., Lang, L., & Yarker, S. (2023). Precarity and the pandemic: The impact of COVID-19 on single men living alone. *The Gerontologist, 63*(1), 131–139. https://doi.org/10.1093/geront/gnac129

Li, M., Li, C., & Zhang, M. (2018). Exploring the spatial spillover effects of industrialization and urbanization factors on pollutants emissions in China's Huang-Huai-Hai region. *Journal of Cleaner Production, 195*, 154–162. https://doi.org/10.1016/j.jclepro.2018.05.186

Lim, W. M., Chin, M. W. C., Ee, Y. S., Fung, C. Y., Giang, C. S., Heng, K. S., Kong, M. L. F., Lim, A. S. S., Lim, B. C. Y., Lim, R. T. H., Lim, T. Y., Ling, C. C., Mandrinos, S., Nwobodo, S., Phang, C. S. C., She, L., Sim, C. H., Su, S. I., Wee, G. W. E., & Weissmann, M. A. (2022). What is at stake in a war? A prospective evaluation of the Ukraine and Russia conflict for business and society. *Global Business and Organizational Excellence, 1–14*. https://doi.org/10.1002/joe.22162

Löckenhoff, C. E., & Carstensen, L. L. (2004). Socioemotional selectivity theory, aging, and health: The increasingly delicate balance between regulating emotions and making tough choices. *Journal of Personality, 72*(6), 1395–1424. https://doi.org/10.1111/j.1467-6494.2004.00301.x

Mai, F., Del Pinto, R., & Ferri, C. (2020). COVID-19 and cardiovascular diseases. *Journal of Cardiology, 76*(5), 453–458. https://doi.org/10.1016/j.jjcc.2020.07.013

Malik, S., Lee, D. C., Doran, K. M., Grudzen, C. R., Worthing, J., Portelli, I., Goldfrank, L. R., & Smith, S. W. (2018). Vulnerability of older adults in disasters: Emergency department utilization by geriatric patients after hurricane sandy. *Disaster Medicine and Public Health Preparedness, 12*(2), 184–193. https://doi.org/10.1017/dmp.2017.44

McCurry, J. (2015). Japan will be model for future super-ageing societies. *The Lancet, 386*(10003), 1523. https://doi.org/10.1016/S0140-6736(15)00525-5

Muramatsu, N., & Akiyama, H. (2011). Japan: Super-aging society preparing for the future. *Gerontologist, 51*(4), 425–432. https://doi.org/10.1093/geront/gnr067

Musich, S., Wang, S. S., Schaeffer, J. A., Kraemer, S., Wicker, E., & Yeh, C. S. (2022). The association of increasing resilience with positive health outcomes among older adults. *Geriatric Nursing, 44*, 97–104. https://doi.org/10.1016/j.gerinurse.2022.01.007

Neves, B. B., Colón Cabrera, D., Sanders, A., & Warren, N. (2023). Pandemic diaries: Lived experiences of loneliness, loss, and hope among older adults during COVID-19. *The Gerontologist, 63*(1), 120–130. https://doi.org/10.1093/geront/gnac104

Oberoi, S., Devleesschauwer, B., Gibb, H. J., & Barchowsky, A. (2019). Global burden of cancer and coronary heart disease resulting from dietary exposure to arsenic, 2015. *Environmental Research, 171*(January), 185–192. https://doi.org/10.1016/j.envres.2019.01.025

Oleksiyenko, A., Terepyshchyi, S., Gomilko, O., & Svyrydenko, D. (2021). 'What do you mean, you are a refugee in your own country?': Displaced scholars and identities in embattled Ukraine. *European Journal of Higher Education, 11*(2), 101–118. https://doi.org/10.1080/21568235.2020.1777446

Onapajo, H. (2014). *Africa spectrum violence and votes in Nigeria* (pp. 27–51).

Pani-Harreman, K. E., Bours, G. J. J. W., Zander, I., Kempen, G. I. J. M., & Van Duren, J. M. A. (2020). Definitions, key themes and aspects of "ageing in place": A scoping review. *Ageing and Society*, 1–34. https://doi.org/10.1017/S0144686X20000094

Pani-Harreman, K. E., Bours, G. J. J. W., Zander, I., Kempen, G. I. J. M., & Van Duren, J. M. A. (2021). Definitions, key themes and aspects of "ageing in place": A scoping review. *Ageing and Society, 41*(9), 2026–2059. https://doi.org/10.1017/S0144686X20000094

Peel, N., Bartlett, H., & McClure, R. (2004). Healthy ageing: How is it defined and measured? *Australasian Journal on Ageing, 23*(3), 115–119. https://doi.org/10.1111/j.1741-6612.2004.00035.x

Quinn, V. J. M., Dhabalia, T. J., Roslycky, L. L., Wilson, V. J. M., Hansen, J. C., Hulchiy, O., Golubovskaya, O., Buriachyk, M., Vadim, K., Zauralskyy, R., Vyrva, O., Stepanskyi, D., Ivanovitch, P. S., Mironenko, A., Shportko, V., & McElligott, J. E. (2022). COVID-19 at war: The Joint Forces Operation in Ukraine. *Disaster Medicine and Public Health Preparedness, 16*(5), 1753–1760. https://doi.org/10.1017/dmp.2021.88

Ross, J. (2020). Economic ramifications of the COVID-19 pandemic for higher education: A circuit breaker in Australian universities' business model? *Higher Education Research and Development, 39*(7), 1351–1356. https://doi.org/10.1080/07294360.2020.1825350

Rudnicka, E., Napierała, P., Podfigurna, A., Męczekalski, B., Smolarczyk, R., & Grymowicz, M. (2020). The World Health Organization (WHO) approach to healthy ageing. *Maturitas, 139*(February), 6–11. https://doi.org/10.1016/j.maturitas.2020.05.018

Sallis, J. F., Kerr, J., Carlson, J. A., Norman, G. J., Saelens, B. E., Durant, N., & Ainsworth, B. E. (2010). Evaluating a brief self-report measure of neighborhood environments for physical activity research and surveillance: Physical Activity Neighborhood Environment Scale (PANES). *Journal of Physical Activity and Health, 7*(4), 533–540. https://doi.org/10.1123/jpah.7.4.533

Sezgin, D., O'Donovan, M., Cornally, N., Liew, A., & O'Caoimh, R. (2019). Defining frailty for healthcare practice and research: A qualitative systematic review with thematic analysis. *International Journal of Nursing Studies, 92*, 16–26. https://doi.org/10.1016/j.ijnurstu.2018.12.014

Streiner, D. L., & Kottner, J. (2014). Recommendations for reporting the results of studies of instrument and scale development and testing. *Journal of Advanced Nursing, 70*(9), 1970–1979. https://doi.org/10.1111/jan.12402

Strong, J., Varady, C., Chahda, N., Doocy, S., & Burnham, G. (2015). Health status and health needs of older refugees from Syria in Lebanon. *Conflict and Health, 9*(1), 1–10. https://doi.org/10.1186/s13031-014-0029-y

Ullah, A. A., Nawaz, F., & Chattoraj, D. (2021). Locked up under lockdown: The COVID-19 pandemic and the migrant population. *Social Sciences & Humanities Open, 3*(1), 100126. https://doi.org/10.1016/j.ssaho.2021.100126

Wahl, H. W., & Gerstorf, D. (2018). A conceptual framework for studying COntext Dynamics in Aging (CODA). *Developmental Review, 50*(August), 155–176. https://doi.org/10.1016/j.dr.2018.09.003

Wahl, H. W., & Gerstorf, D. (2020). Person-environment resources for aging well: Environmental docility and life space as conceptual pillars for future contextual gerontology. *Gerontologist, 60*(3), 368–375. https://doi.org/10.1093/geront/gnaa006

World Health Organization. (2019). *World population ageing 2019*. http://link.springer.com/chapter/10.1007/978-94-007-5204-7_6

Chapter 3
The Impact of Crises on Older Adults' Health and Function: An Intergenerational Perspective

Whitney Nesser and Thomas W. Nesser

3.1 Introduction

As the global population continues to age, it is important to understand how crises such as the coronavirus pandemic, climate change, and violence impact older adults' health and functional ability. Ageing is often accompanied by a decline in physical, cognitive, and social abilities (Bernabe et al., 2023; Mogic et al., 2023; Robertson et al., 2023), directly affecting the ability to function without assistance and lowering quality of life. This decline in function can also have broader personal, social, and economic implications. Research has identified strategies and interventions to help older adults maintain or improve their health and function, but there is limited information on the impact of global crises on health for those ageing in place in different generations (Ayalon et al., 2022; Barchielli et al., 2022; Kennedy et al., 2023).

Generations are typically defined as a cohort of individuals who were born at the same time and have similar lived experiences. For this chapter, the discussion surrounding generations will mainly focus on older adults in respective age-defined decades (i.e. 50–59) and the intergenerational differences between decades. While these age-defined decades are not absolute with similar health profiles, research supports that age groups differ, particularly when the span of age is broad, such as midlife and old age (Wettstein et al., 2020). One study found that certain personality traits and attitudes towards one's own ageing may impact, and perhaps predict, objective health changes in later life (Wettstein et al., 2020).

W. Nesser (✉)
Department of Applied Clinical and Educational Sciences, Indiana State University, Terre Haute, IN, USA
e-mail: whitney.nesser@indstate.edu

T. W. Nesser
Department of Kinesiology, Recreation, and Sport, Indiana State University, Terre Haute, IN, USA

N. Asiamah et al. (eds.), *Sustainable Neighbourhoods for Ageing in Place*,
https://doi.org/10.1007/978-3-031-41594-4_3

Older adults within different age groups have varying health profiles, particularly for those who are ageing in place without convenient resources to maintain good health. Worldwide, an individual's health status depends on a number of elements, including healthcare accessibility, government policies, personality traits, communicable and noncommunicable diseases, disability, and socioeconomic status (Kampfen et al., 2018; Wettstein et al., 2020). Three of the most confounding factors determining health status in generations of older adults ageing in place is response to the crises: coronavirus pandemic, climate change, and violence (Ayalon et al., 2022; Barchielli et al., 2022; Danielson & Ray-Degges, 2022; Kennedy et al., 2023). While there are various ways to measure health, the most comprehensive measurement is to assess multiple domains of health. These domains can include physical health, psychological health, social interaction, and many others. By measuring different health domains, it allows a more comprehensive picture of health and tailor interventions and resources to specific ageing populations (Nesser et al., 2023).

This chapter comprises seven main sections. The first section (i.e. the introduction) provides the general background of the chapter, with the second section providing a theoretical framework to support the multiple influences involved in maintaining health for those ageing in place within the context of three crises. The third section explains health profile measurements to best assess well-being for older adults ageing in place. In the fourth section, resilience and autonomy are discussed from an intergenerational perspective. The fifth section explores implications for ageing and policy within these crises. We discuss limitations and future research in the sixth section, and to conclude the chapter, we reflect on ways to further support healthy ageing in place during global crises. To provide context to how healthy ageing in place is impacted by the crises, we provide an overview of the ecological theoretical framework.

3.2 Theoretical Framework

Ecological models of health have developed from public health and psychology. In public health, the host-agent-environment model is basic to infectious disease analysis but can also be applied to chronic diseases (McLeroy et al., 1988). The work of Bronfenbrenner forms the basis of health ecological models (Bronfenbrenner, 1979; McLeroy et al., 1988). Four core assumptions form the ecological perspective in the field of health. These four core assumptions frame the ecological levels of analysis (Stokols, 1992).

The first assumption in the ecological model is that personal well-being includes genetics, attitudes, and behaviors (Stokols, 1992). Health profiles encompass multiple facets of one's individuality and environment. This is particularly important when recognizing the environmental factors related to crises impacting older adults ageing in place. For example, multiple generations of older adults have reported fear related to energy cost inflation (Barchielli et al., 2022), and homicidal violence among older adults occurs primarily at their personal dwelling and much more often than younger persons (Kennedy et al., 2023). The second assumption is that the analysis of health must take into consideration multiple aspects related to where one lives (Stokols, 1992). Related to this second assumption is the impact on health living conditions created by global crises. For example, declining health occurs for those ageing in place as a result of increasing global temperatures, lack of healthcare access during the pandemic, and living environments with little to no protection from violence (Ayalon et al., 2022; Barchielli et al., 2022; Kennedy et al., 2023). The third assumption is that the effectiveness of maintaining functional ability is enhanced through individual and group efforts and global health policies that include health metrics of at-risk older adults (Green & Kreuter, 1999; Malkowski et al., 2023). Interpersonal and institutional aspects of the ecological perspective are particularly relevant to intergenerational older adults with varying health profiles. Open communication between family members and support from friends and neighbors improve health for those ageing in place (Lowers et al., 2023). The fourth assumption is that reciprocity exists between individuals and their environment. The physical and social aspects of autonomous living and resilience directly influence an individual's health (Bhat et al., 2022). Similarly, health is modified by an individual through their surroundings with individual and collective actions (Salnikov & Baramiya, 2021).

The ecological perspective is employed to examine intrapersonal, interpersonal, and institutional factors in various populations and studies. Studies have explored the ecological model as a framework for food insecurity (Freiria et al., 2022). A scoping review of food insecurity among older adults utilized the ecological model to summarize the results. In this review, the majority of studies found focused on intrapersonal factors as the main relationship to food insecurity (Freiria et al., 2022). Additionally, a multitude of studies have found food insecurity to be associated with the coronavirus pandemic among older adults (Alan et al., 2023; Cai & Bidulescu, 2023; Nicklett et al., 2023).

Within the ecological framework is the complex interplay of factors, including those influencing health outcomes for older adults at multiple levels. For example, engaging in regular exercise and maintaining a healthy diet can help prevent or manage chronic conditions such as diabetes and heart disease (Cadore & Izquierdo, 2015). However, this becomes a particular challenge related to anxiety and mental health when ageing in place occurs in unsafe communities with limited healthcare access (Golovchanova et al., 2023). At the social level, social support, access to healthcare, and community resources can impact health outcomes for older adults (Asiamah et al., 2023). In the combined social and environmental levels for older adults, factors such as neighbourhood safety, social support provided, and walkability can improve health (Asiamah et al., 2022). For example, living in a neighbourhood with safe sidewalks and accessible public transportation can help older adults stay physically active and engaged in their communities (Hirsch et al., 2022; Siqueira Junior et al., 2022). It is critical older adults are safe and protected from crime and violence in their communities to stay physically active and engaged (Kennedy et al., 2023). Related to health, the ecological framework when applied to older adults recognizes these factors are interconnected and influence health outcomes in complex ways. Interventions promoting social support and community engagement can help address social isolation or loneliness, and policies prioritizing safe and accessible transportation can help older adults maintain their mobility and independence. This theoretical framework provides context for examining health, especially when contextualizing health within crises such as the coronavirus pandemic, climate change, and violence. In addition to the multilayered ecological framework, it is important to select appropriate health profile instruments to accurately measure aspects of health such as functional ability and quality of life.

3.3 Functional Ability as Measured by Quality of Life Profiles

Functional ability is defined as the ability to accomplish routine tasks and activities required for independent living. These activities include activities of daily living (ADLs) and instrumental activities of daily living (IADLs) (Zhu et al., 2023). ADLs are basic self-care tasks, such as hygiene, eating, and walking. IADLs are more complex tasks, such as caring for others, driving, and managing finances. (Pashmdarfard & Azad, 2020; Zhu et al., 2023). Functional ability is a critical aspect of healthy ageing. As functional ability declines in older adults, their ability to live without assistance, engage in leisure and social activities reduces, and quality of life reduces. Factors contributing to declines in functional ability include chronic conditions, such as arthritis or heart disease, cognitive impairment, mobility limitations, and limited social support. These factors are also directly attributable to crises impacting the health of older adults ageing in place.

However, with appropriate interventions and support, functional ability can often be maintained or improved in older adults (Sharma et al., 2021; Sheshadri & Elia, 2022). Examples of interventions that can improve functional ability include physical activity promotion, physical rehabilitation, supporting older adults to use assistive devices, and modifications to the living environment. Overall, functional ability is a key component of healthy ageing and is an important consideration in healthcare and public health efforts aimed at promoting the well-being of older adults.

Social and emotional health are interconnected, and older adults who experience social isolation or loneliness may be at higher risk for depression, anxiety, and other mental health problems (Dugan & Kivett, 1994; Lemish, 2020). Similarly, older adults who experience emotional stress or trauma may be at higher risk for social isolation and loneliness. Cognitive and physical health are also closely linked, and older adults who experience cognitive decline may be at higher risk for physical health problems due to a lack of engagement in activities that promote physical health, such as exercise and social engagement (Handing et al., 2023). Climate change and environmental factors, such as air pollution or exposure to toxins, can also impact the physical health of older adults, particularly those with underlying health conditions (Fernandes et al., 2021). It is important to take a holistic approach to the health of older adults and consider the interconnectivity of various health domains. By contextualizing what it means to promote healthy behaviors while addressing underlying health conditions within the sphere of crises, it will be possible to improve multiple aspects of health and support overall well-being for those ageing in place.

Quality of life is a multidimensional measure of health that contains the abovementioned domains of health (World Health Organization, 1998). An abundance of general health profile and quality of life instruments are available. Quality of life instruments can be disease- or population-specific with many developed for the general population, while some assess individuals with chronic illness or cancer, and others were developed for caregivers (Martin et al., 2021; Nesser et al., 2023). The more recognizable general population health profiles include measures, such as the Medical Outcomes Short Form 12 item and 36 item scales (Gandek et al., 1998; Ware, 1996) the World Health Organization Quality of Life Brief Version (26 items) and 100 item scales (World Health Organization, 1998). Each of these instruments measure something slightly different, although they all have domains and items to measure physical health and emotional/psychological health. The World Health Organization scale is the most comprehensive assessment for the general population with six quality of life domains (Nesser et al., 2023; World Health Organization, 1998). These domains include physical health (e.g. sleep, pain), psychological health (e.g. memory, feeling positive), level of independence (e.g. ability to work), social relationships (e.g. family, friends), environment (e.g. safety, transportation), and spirituality/religion/personal beliefs (World Health Organization, 2012).

Differences in health begin to emerge during middle and older aged individuals. These differences have been found to be attributed to changes in physical health, cognitive function, and ability to interact socially (Ailshire & Crimmins, 2011; Rozani, 2022). Beginning at age 50, individuals may start to experience age-related changes in vision, hearing, and mobility (Kuo et al., 2022; Luo et al., 2022). By age 60, individuals may start to experience more significant declines in physical function and may be at higher risk for falls and injuries (Aburub et al., 2023; Garbin & Fisher, 2023; Jayakody et al., 2023) while experiencing changes in cognitive function, such as memory loss and decreased processing speed. At age 70, individuals may experience further declines in physical function and mobility and may be at increased risk for chronic conditions, such as dementia and Parkinson's disease. By age 80, individuals may experience significant declines in physical and cognitive function and may require more care. Osteoporosis is also common at this age mostly in females (Aburub et al., 2023). Overall, the risk of chronic conditions and declines in physical and cognitive function have been found to become more prevalent as individuals age. (Franceschi et al., 2018). Individual health outcomes can sometimes be modified, such as increasing physical activity and consuming a healthy diet. One of the main predictors of reducing overall health decline is the ability to be resilient and autonomous.

3.4 Intergenerational Resilience and Autonomy

Intergenerational differences can be a significant issue for older adults, especially as they may feel disconnected from younger generations who may have different cultural values, social norms, and communication styles. Some ways intergenerational differences impact older adults include communication, technology, and values/beliefs. Technological intergenerational differences can be significant (Scarpina et al., 2021). Older adults may struggle with new technology or find it challenging to keep up with the latest devices or software (Dust et al., 2019). For example, older adults may have different attitudes towards issues, such as social justice, gender roles, or religion than younger generations. Despite these challenges, there are also many opportunities for intergenerational connections and understanding. For example, the coronavirus pandemic may have served as a catalyst for helping to bridge some of the intergenerational digital communications between young and old (Vlachantoni et al., 2023). By promoting dialogue and mutual respect, younger and older generations can learn from each other, share experiences, and build relationships that can help bridge the gap between different age groups. Intergenerational programs and activities, such as mentorship programs and community service projects, allow different generations to interact and learn from one another. (Anderson et al., 2017; Turner et al., 2023).

The interconnectivity of quality of life domains (i.e. psychological, social, physical) is notable in all age groups but particularly in those who are entering midlife and beyond. Coping with crises, such as the coronavirus pandemic and violence, further deepens the association among the domains. It has been found that for women aged 52–69, psychological well-being in midlife shows an association with cognitive function in later life (Nakanishi et al., 2019). Sleep duration and continuity has negative effects on physical and psychological aspects of ageing. However, researchers conducting a 12-year longitudinal study on 300 women at ages 52, 55, and 64 (three point-in-time assessments with the same participants) found the inverse result, where sleep characteristics may not become worse with early older ageing (Matthews et al., 2020). All ageing research is not as positive as the Matthews et al. sleep study. Ageing has been found to be inextricably linked to depression, anxiety, loneliness, and comorbid anxiety-depressive symptoms for autonomous community-dwelling individuals between the ages of 60 and 88 (Pedroso-Chaparro et al., 2023). These symptoms are exacerbated by all three of the crises discussed in this chapter. The coronavirus pandemic has increased loneliness in older adults (Morgan et al., 2023), climate change has contributed to food insecurity (Mahmood et al., 2022; Verschuur et al., 2021), and neighborhood violence decreases access to support services (Sheppard et al., 2022).

Middle-aged individuals (e.g. age 50) are still in their prime working years and may be juggling career and family responsibilities. Autonomy and resilience can help maintain a sense of control over their lives and cope with stressors such as job loss or family illness (Yang et al., 2019). Building resilience at this age can also help prevent or mitigate the negative effects of chronic stress on health. By age 60, many individuals are nearing retirement age and may be facing changes in their social roles and routines. Autonomy can help them maintain a sense of purpose and control over their lives as they transition into retirement. Resilience can also be important at this age to help individuals cope with age-related declines in physical and cognitive function (Henry et al., 2023). At age 70, many individuals may be dealing with chronic conditions and age-related health issues. Autonomy can help them maintain their independence and sense of identity despite these challenges. Resilience can also be important to help individuals adapt to changes in physical health and psychological health, while maintaining social connections. By age 80, many individuals may be dealing with significant declines in physical and cognitive function and may require more assistance with activities of daily living. Autonomy can help them maintain a sense of dignity and control over their lives despite these challenges. Resilience can also be important to help individuals cope with losses such as the death of a spouse or changes in living arrangements (Kawas & Corrada, 2020). Building resilience and promoting autonomy helps prevent and mitigate negative health outcomes associated with chronic stress and age-related declines in physical and cognitive function.

Research is lacking in the area of resilience in advanced age, but some researchers have found that for those above 85 years of age, physical, social, and psychological characteristics pose a unique balance contributing to resilience. The longer life is lived, the more the balance becomes one between losses, vulnerability, and lack of resources with wisdom, experience, support systems, and autonomy (Hayman et al., 2017). Using a socio-ecological model of health, Paine et al. conducted qualitative in-depth interviews with adults aged 80 and older in Australia. Findings revealed positive resilience from an intrapersonal perspective despite the challenges of living alone in government housing. Participants also felt a sense of autonomy from living independently (Paine et al., 2022). These results are supported by researchers who studied indigenous older adults living in geographically isolated rural areas in in Chile (Gallardo-Peralta et al., 2023). They found that accessible residential settings encourage social participation, an attribute that can reduce depressive symptoms by community integration (Gallardo-Peralta et al., 2023). Even in the face of crises, such as accelerated violence in neighborhoods, many older adults continue to be resilient by relying on social support (Taei et al., 2022).

One promising method to enhance resilience and autonomy in advanced age is family group conferencing, where meetings are led by and held with family members to empower collective strength in support of the older adult family members by developing care plans. What is of utmost importance in this method however is to ensure the focus is on reciprocity with a relational empowerment strengthening model to support the needs of advanced age family members (Metze et al., 2015a, b). In addition to the psychological and social aspects of advanced age, the physical aspect of health profiles has significant challenges for most individuals. In a study of 961 community-dwelling persons aged 75–85 years of age, researchers found that resilience moderated walking difficulty for some, but not among those 85 years of age (Siltanen et al., 2021). Once severe walking or physical health limitations occur, resilience does not appear to be a mitigating factor to reduce the severity (Siltanen et al., 2021). However, establishing policy can serve as a mitigating factor to avoid negative impacts of the crises on access to essential amenities, such as quality healthcare, nutritious foods, social support, and safe housing.

3.5 Implications for Ageing and Policy

Access to essential amenities is negatively impacted by crises, such as the coronavirus pandemic, climate change, and violence. However, research has shown there are ways to reduce or avoid the negative crises impact on access to these amenities. The coronavirus pandemic had significant impact on older adults; however, related vaccine policies began to reduce the impact of the transmission and infection in older adults who are ageing in place (Arthur-Holmes et al., 2020; Liang et al., 2022; Medina-Walpole, 2020).

Climate change has far-reaching impact on the health of older adults who are more susceptible to extreme heat and pollution (Neira et al., 2023). Yet, research suggests that with climate change policies, such as evaluating water quality with conjoined efforts for increasing access to clean water, the health of older adults improve, further reducing the impact of the crises (Neira et al., 2023). When the population that is directly impacted by climate change are consulted and involved in the creation of policy to mitigate the crises, extreme aspects of the crises can averted (Martel-Morin & Lachapelle, 2022).

Kennedy et al. conducted a meta-analysis with results supporting the need for research and policy to focus on how violence impacts older adults differently than other generations (Kennedy et al., 2023). Highlighting the unique needs of those who are ageing in place allows for tailored prevention efforts to help avoid negative crises impact on social support and neighborhood safety (Kennedy et al., 2023).

There is an inextricable connection among crises, and the coronavirus pandemic, climate change, and violence are no exception. Separating the crises to identify the health impacts on older adults is feasible. However, when policies are made to avoid the negative impact of the crises on health and essential amenities, they must all be created specifically with older adults who are ageing in place as a unique group and be locally and regionally appropriate.

3.6 Limitations and Future Research

We have provided an overview of older adults who are ageing in place during various life stages within the context of three global crises: the coronavirus pandemic, climate change, and violence. Health profiles are dependent upon certain variables and aspects, such as genetics, modifiable health behaviors, socioeconomic status, and others.

Our premise is that when available, preventive services supersede secondary and tertiary healthcare. However, this chapter does not include specific evidence to support all conditions associated with health in each older adult age group, nor does it include all global crises. To have a precise determination of resilience and autonomy as they relate to each crisis and are represented in health profiles, research tailored to each group should include some essential components. Future researchers are encouraged to select relevant health profile instruments adapted to the population being studied (i.e. those in different age ranges, the variability of neighbourhoods, subjective perceptions on ageing in place).

Future research can span a range of topics to better understand the complex interplay of factors that impact health outcomes for the ageing population. Some examples include longitudinal studies, multidisciplinary research, interventions, and advanced technology. Longitudinal studies that follow individuals over time can provide valuable insights into how health profiles change with age and what factors contribute to these changes. Such studies can help identify early indicators of health problems and inform interventions to prevent or mitigate negative health outcomes that are influenced by a range of factors at the individual, social, and environmental levels. Multidisciplinary perspectives from experts in different fields can advance future ageing research. Future research could focus on developing and testing interventions to improve health outcomes for older adults. These interventions could target a range of factors, including lifestyle behaviors, social support, and access to healthcare and community resources. Advances in technology have the potential to improve health outcomes for older adults, for example, through the development of wearable devices and telehealth technologies. Future research could focus on the use of technology to monitor and improve health outcomes for older adults, as well as to enhance social connectedness and quality of life. It is recommended that future researchers investigate all WHO quality of life domains to garner a more comprehensive picture of health for tailored interventions and resources in specific ageing populations.

3.7 Conclusions

The impact of global crises such as the coronavirus pandemic, climate change, and violence on older adults' health and function extends beyond individual well-being to broader societal implications. The decline in physical, cognitive, and social abilities that often accompanies ageing can lead to increased healthcare costs and the need for caregiving services. However, by promoting healthy ageing through interventions, such as exercise, assistive devices, home modifications, and cognitive interventions, we can support older adults' independence, autonomy, and resiliency as they face challenges compounded by global crises. It is important to continue prioritizing research and intervention efforts aimed at improving older adults' health and function, recognizing the impact on health by global crises. By investing in healthy ageing and being mindful of the world in which we live, we can help ensure that older adults are able to age with dignity, independence, and a high quality of life.

Review Questions

1. How would you define functional ability and health within the context of ageing in place and global crises? What are a few ways to measure health status in older adults?
2. What are some intergenerational differences between and among the health profiles of older adults who are 'ageing in place' within the context of global crises?
3. Which of the World Health Organizational health domains do you feel are important for supporting the functional ability and health for 'ageing in place' as older adults are impacted by the global crises?
4. How would you describe ways in which the personal, community, and environmental factors of the ecological model influence functional ability and health of older adults in different age groups as they contend with the global crises?
5. Based on this chapter, what are some innovative programming ideas to improve the health profiles of older adults and mitigate the global crises affecting sustainable neighbourhoods?

References

Aburub, A. S., Phillips, S. P., Curcio, C. L., Guerra, R. O., Khalil, H., & Auais, M. (2023). Circumstances and factors associated with falls among community-dwelling older adults diagnosed with heart disease using the International Mobility in Aging Study (IMIAS). *Journal of Geriatric Physical Therapy (2001), 46*(1), 53–63. https://doi.org/10.1519/JPT.0000000000000316

Ailshire, J. A., & Crimmins, E. M. (2011). Psychosocial factors associated with longevity in the United States: Age differences between the old and oldest-old in the health and retirement study. *Journal of Aging Research, 2011*, 530534. https://doi.org/10.4061/2011/530534

Alan, V. M., Oscar, R. C., Pablo, G. R., Monica, A. M., & Miriam, L. T. (2023). Prevalence of food insecurity and its association with depressive and anxiety symptoms in older adults during the COVID-19 pandemic in Mexico: A secondary analysis of ENCOVID-19 survey. *Frontiers in Medicine (Lausanne), 10*, 1110584. https://doi.org/10.3389/fmed.2023.1110584

Anderson, S., Fast, J., Keating, N., Eales, J., Chivers, S., & Barnet, D. (2017). Translating knowledge: Promoting health through intergenerational community arts programming. *Health Promotion Practice, 18*(1), 15–25. https://doi.org/10.1177/1524839915625037

Arthur-Holmes, F., Akaadom, M. K. A., Agyemang-Duah, W., Abrefa Busia, K., & Peprah, P. (2020). Healthcare concerns of older adults during the COVID-19 outbreak in low- and middle-income countries: Lessons for health policy and social work. *Journal of Gerontological Social Work, 63*(6-7), 717–723. https://doi.org/10.1080/01634372.2020.1800883

Asiamah, N., Lowry, R., Khan, H. T. A., & Awuviry-Newton, K. (2022). Associations between social support provided and walkability among older adults: Health self-consciousness as a moderator. *Archives of Gerontology and Geriatrics, 101*, 104691. https://doi.org/10.1016/j.archger.2022.104691

Asiamah, N., Kouveliotis, K., Eduafo, R., & Borkey, R. (2023). Associations between vigorous physical activity, social ties, social support, and self-reported health among older adults in Accra, Ghana. *PLOS Global Public Health, 3*(2), e0001582. https://doi.org/10.1371/journal.pgph.0001582

Ayalon, L., Roy, S., Aloni, O., & Keating, N. (2022). A scoping review of research on older people and intergenerational relations in the context of climate change. *The Gerontologist*. https://doi.org/10.1093/geront/gnac028

Barchielli, B., Cricenti, C., Galle, F., Sabella, E. A., Liguori, F., Da Molin, G., et al. (2022). Climate changes, natural resources depletion, COVID-19 pandemic, and Russian-Ukrainian War: What is the impact on habits change and mental health? *International Journal of Environmental Research and Public Health, 19*(19). https://doi.org/10.3390/ijerph191911929

Bernabe, E., de Oliveira, C., de Oliveira Duarte, Y. A., Bof de Andrade, F., & Sabbah, W. (2023). Social participation and tooth loss, vision, and hearing impairments among older Brazilian adults. *Journal of the American Geriatrics Society*. https://doi.org/10.1111/jgs.18423

Bhat, A. C., Almeida, D. M., Fenelon, A., & Santos-Lozada, A. R. (2022). A longitudinal analysis of the relationship between housing insecurity and physical health among midlife and aging adults in the United States. *SSM – Population Health, 18*, 101128. https://doi.org/10.1016/j.ssmph.2022.101128

Bronfenbrenner, U. (1979). *The ecology of human development: Experiments by nature and design*. Harvard University Press.

Cadore, E. L., & Izquierdo, M. (2015). Exercise interventions in polypathological aging patients that coexist with diabetes mellitus: Improving functional status and quality of life. *Age (Dordrecht, Netherlands), 37*(3), 64. https://doi.org/10.1007/s11357-015-9800-2

Cai, J., & Bidulescu, A. (2023). The association between chronic conditions, COVID-19 infection, and food insecurity among the older US adults: Findings from the 2020-2021 National Health Interview Survey. *BMC Public Health, 23*(1), 179. https://doi.org/10.1186/s12889-023-15061-8

Danielson, R. A., & Ray-Degges, S. (2022). Aging in place among older adults with histories of traumatic experiences: A scoping review. *Gerontologist, 62*(1), e1–e16. https://doi.org/10.1093/geront/gnab127

Dugan, E., & Kivett, V. R. (1994). The importance of emotional and social isolation to loneliness among very old rural adults. *Gerontologist, 34*(3), 340–346. https://doi.org/10.1093/geront/34.3.340

Dust, S. B., Gerhardt, M. W., Hebbalalu, D., & Murray, M. (2019). Protecting my turf: The moderating role of generational differences on the relationships between self-direction and hedonism values and reactions to generational diversity. *The Journal of Social Psychology, 159*(2), 153–169. https://doi.org/10.1080/00224545.2019.1570903

Fernandes, A., Van Lenthe, F. J., Vallee, J., Sueur, C., & Chaix, B. (2021). Linking physical and social environments with mental health in old age: A multisensor approach for continuous real-life ecological and emotional assessment. *Journal of Epidemiology and Community Health, 75*(5), 477–483. https://doi.org/10.1136/jech-2020-214274

Franceschi, C., Garagnani, P., Morsiani, C., Conte, M., Santoro, A., Grignolio, A., et al. (2018). The continuum of aging and age-related diseases: Common mechanisms but different rates. *Frontiers in Medicine (Lausanne), 5*, 61. https://doi.org/10.3389/fmed.2018.00061

Freiria, C. N., Arikawa, A., Van Horn, L. T., Corona, L. P., & Wright, L. Y. (2022). Food insecurity among older adults living in low- and middle-income countries: A scoping review. *Gerontologist*. https://doi.org/10.1093/geront/gnac161

Gallardo-Peralta, L. P., Raymond, E., & Galvez-Nieto, J. L. (2023). Ageing in context: An ecological model to understand social participation among indigenous adults in Chile. *Research on Aging, 45*(3-4), 332–346. https://doi.org/10.1177/01640275221108502

Gandek, B., Ware, J. E., Aaronson, N. K., Apolone, G., Bjorner, J. B., Brazier, J. E., et al. (1998). Cross-validation of item selection and scoring for the SF-12 Health Survey in nine countries: Results from the IQOLA Project. International Quality of Life Assessment. *Journal of Clinical Epidemiology, 51*(11), 1171–1178. https://doi.org/10.1016/s0895-4356(98)00109-7

Garbin, A. J., & Fisher, B. E. (2023). The interplay between fear of falling, balance performance, and future falls: Data from the National Health and Aging Trends Study. *Journal of Geriatric Physical Therapy (2001), 46*(2), 110–115. https://doi.org/10.1519/JPT.0000000000000324

Golovchanova, N., Evans, B., Hellfeldt, K., Andershed, H., & Boersma, K. (2023). Older and feeling unsafe? Differences in underlying vulnerability, anxiety and life satisfaction among older adults. *Aging & Mental Health*, 1–8. https://doi.org/10.1080/13607863.2023.2177255

Green, L. W., & Kreuter, M. W. (1999). *Health promotion planning: An educational and ecological approach* (3rd ed.). Mayfield Pub. Co..

Handing, E. P., Hayden, K. M., Leng, X. I., & Kritchevsky, S. B. (2023). Predictors of cognitive and physical decline: Results from the Health Aging and Body Composition Study. *Frontiers in Aging Neuroscience, 15*, 1122421. https://doi.org/10.3389/fnagi.2023.1122421

Hayman, K. J., Kerse, N., & Consedine, N. S. (2017). Resilience in context: The special case of advanced age. *Aging & Mental Health, 21*(6), 577–585. https://doi.org/10.1080/13607863.2016.1196336

Henry, J. D., Grainger, S. A., & von Hippel, W. (2023). Determinants of social cognitive aging: Predicting resilience and risk. *Annual Review of Psychology, 74*, 167–192. https://doi.org/10.1146/annurev-psych-033020-121832

Hirsch, J. A., Michael, Y. L., Moore, K. A., Melly, S., Hughes, T. M., Hayden, K., et al. (2022). Longitudinal neighbourhood determinants with cognitive health and dementia disparities: Protocol of the Multi-Ethnic Study of Atherosclerosis Neighborhoods and Aging prospective cohort study. *BMJ Open, 12*(11), e066971. https://doi.org/10.1136/bmjopen-2022-066971

Jayakody, O., Blumen, H. M., Ayers, E., & Verghese, J. (2023). Longitudinal associations between falls and risk of gait decline: Results from the central control of mobility and aging study. *Archives of Physical Medicine and Rehabilitation, 104*(2), 245–250. https://doi.org/10.1016/j.apmr.2022.08.975

Kampfen, F., Wijemunige, N., & Evangelista, B., Jr. (2018). Aging, non-communicable diseases, and old-age disability in low- and middle-income countries: A challenge for global health. *International Journal of Public Health, 63*(9), 1011–1012. https://doi.org/10.1007/s00038-018-1137-z

Kawas, C. H., & Corrada, M. M. (2020). Successful cognitive aging: What the oldest-old can teach us about resistance and resilience. *Neurology, 95*(8), 329–330. https://doi.org/10.1212/WNL.0000000000010251

Kennedy, B., Bugeja, L., Olivier, J., Johnson, M., Hua, P., Koppel, S., & Ibrahim, J. E. (2023). Epidemiology of homicide in community-dwelling older adults: A systematic review and meta-analysis. *Trauma Violence Abuse, 24*(2), 390–406. https://doi.org/10.1177/15248380211030250

Kuo, F. L., Yen, C. M., Chen, H. J., Liao, Z. Y., & Lee, Y. (2022). Trajectories of mobility difficulty and falls in community-dwelling adults aged 50 + in Taiwan from 2003 to 2015. *BMC Geriatrics, 22*(1), 902. https://doi.org/10.1186/s12877-022-03613-3

Lemish, V. V. (2020). [Representation of social workers with various professional experience about emotional features of elderly and old people]. *Advances in Gerontology, 33*(5), 893–899. Retrieved from https://www.ncbi.nlm.nih.gov/pubmed/33550744

Liang, C. K., Lee, W. J., Peng, L. N., Meng, L. C., Hsiao, F. Y., & Chen, L. K. (2022). COVID-19 vaccines in older adults: challenges in vaccine development and policy making. *Clinics in Geriatric Medicine, 38*(3), 605–620. https://doi.org/10.1016/j.cger.2022.03.006

Lowers, J., Zhao, D., Bollens-Lund, E., Kavalieratos, D., & Ornstein, K. A. (2023). Solo but not alone: An examination of social and help networks among community-dwelling older adults without close family. *Journal of Applied Gerontology, 42*(3), 419–426. https://doi.org/10.1177/07334648221135588

Luo, J., Agboola, F., Grant, E., Morris, J. C., Masters, C. L., Albert, M. S., et al. (2022). Accelerated longitudinal changes and ordering of Alzheimer disease biomarkers across the adult lifespan. *Brain, 145*(12), 4459–4473. https://doi.org/10.1093/brain/awac238

Mahmood, J., Rajaram, N. N., & Guinto, R. R. (2022). Addressing food insecurity and climate change in Malaysia: Current evidence and ways forward. *The Malaysian Journal of Medical Sciences, 29*(6), 1–5. https://doi.org/10.21315/mjms2022.29.6.1

Malkowski, O. S., Kanabar, R., & Western, M. J. (2023). Socio-economic status and trajectories of a novel multidimensional metric of Active and Healthy Ageing: The English Longitudinal Study of Ageing. *Scientific Reports, 13*(1), 6107. https://doi.org/10.1038/s41598-023-33371-0

Martel-Morin, M., & Lachapelle, E. (2022). The Five Canadas of Climate Change: Using audience segmentation to inform communication on climate policy. *PLoS One, 17*(11), e0273977. https://doi.org/10.1371/journal.pone.0273977

Martin, M. P., McEntee, M. L., & Suri, Y. (2021). Caregiver quality of life: How to measure it and why. *American Journal of Health Promotion, 35*(7), 1042–1045. https://doi.org/10.1177/08901171211030142f

Matthews, K. A., Kravitz, H. M., Lee, L., Harlow, S. D., Bromberger, J. T., Joffe, H., & Hall, M. H. (2020). Does midlife aging impact women's sleep duration, continuity, and timing? A longitudinal analysis from the Study of Women's Health Across the Nation. *Sleep, 43*(4). https://doi.org/10.1093/sleep/zsz259

McLeroy, K. R., Bibeau, D., Steckler, A., & Glanz, K. (1988). An ecological perspective on health promotion programs. *Health Education Quarterly, 15*(4), 351–377. https://doi.org/10.1177/109019818801500401

Medina-Walpole, A. (2020). AGS COVID-19 policy brief offers roadmap to government action for a critical group during pandemics: Older adults in nursing homes. *Journal of Gerontological Nursing, 46*(7), 55–56. https://doi.org/10.3928/00989134-20200614-01

Metze, R. N., Kwekkeboom, R. H., & Abma, T. A. (2015a). The potential of Family Group Conferencing for the resilience and relational autonomy of older adults. *Journal of Aging Studies, 34*, 68–81. https://doi.org/10.1016/j.jaging.2015.04.005

Metze, R. N., Kwekkeboom, R. H., & Abma, T. A. (2015b). 'You don't show everyone your weakness': Older adults' views on using Family Group Conferencing to regain control and autonomy. *Journal of Aging Studies, 34*, 57–67. https://doi.org/10.1016/j.jaging.2015.04.003

Mogic, L., Rutter, E. C., Tyas, S. L., Maxwell, C. J., O'Connell, M. E., & Oremus, M. (2023). Functional social support and cognitive function in middle- and older-aged adults: A systematic review of cross-sectional and cohort studies. *Systematic Reviews, 12*(1), 86. https://doi.org/10.1186/s13643-023-02251-z

Morgan, T., Wiles, J., Morgan, K., Williams, L., Black, S., Koh, A., et al. (2023). Older people's views on loneliness during COVID-19 lockdowns. *Aging & Mental Health*, 1–9. https://doi.org/10.1080/13607863.2023.2211549

Nakanishi, M., Yamasaki, S., Nishida, A., & Richards, M. (2019). Midlife psychological well-being and its impact on cognitive functioning later in life: An observational study using a female British birth cohort. *Journal of Alzheimer's Disease, 72*(3), 835–843. https://doi.org/10.3233/JAD-190590

Neira, M., Erguler, K., Ahmady-Birgani, H., Al-Hmoud, N. D., Fears, R., Gogos, C., et al. (2023). Climate change and human health in the Eastern Mediterranean and Middle East: Literature review, research priorities and policy suggestions. *Environmental Research, 216*(Pt 2), 114537. https://doi.org/10.1016/j.envres.2022.114537

Nesser, W., Yi, E.-H., Wotring, A., Hutchins, M., & Snyder, S. (2023). A partial picture, without a frame: Measuring quality of life in Cystic Fibrosis caregivers. *American Journal of Health Education* (Online, published ahead of print). https://doi.org/10.1080/19325037.2023.2186984

Nicklett, E. J., Cheng, G. J., & Morris, Z. A. (2023). Predictors of food insecurity among older adults before and during COVID-19 in the United States. *Frontiers in Public Health, 11*, 1112575. https://doi.org/10.3389/fpubh.2023.1112575

Paine, N., Lowe, M., Rachele, J., & Turrell, G. (2022). "I'm definitely not socially excluded!" – Perceptions of social exclusion among Australian government housing residents aged 80 and older who live alone. *Journal of Aging Studies, 61*, 101011. https://doi.org/10.1016/j.jaging.2022.101011

Pashmdarfard, M., & Azad, A. (2020). Assessment tools to evaluate Activities of Daily Living (ADL) and Instrumental Activities of Daily Living (IADL) in older adults: A systematic review. *Medical Journal of the Islamic Republic of Iran, 34*, 33. https://doi.org/10.34171/mjiri.34.33

Pedroso-Chaparro, M. D. S., Anton-Lopez, J. C., Cabrera, I., Marquez-Gonzalez, M., Martinez-Huertas, J. A., & Losada-Baltar, A. (2023). 'I feel old and have aging stereotypes'. Internalized aging stereotypes and older adults' mental health: The mediational role of loneliness. *Aging & Mental Health*, 1–8. https://doi.org/10.1080/13607863.2022.2163376

Robertson, M. C., Downer, B., Schulz, P. E., Samper-Ternent, R., Lyons, E. J., & Milani, S. A. (2023). Social and leisure activities predict transitions in cognitive functioning in older Mexican adults: A latent transition analysis of the Mexican Health and Aging Study. *The Journals of Gerontology. Series B, Psychological Sciences and Social Sciences*. https://doi.org/10.1093/geronb/gbad082

Rozani, V. (2022). Ethnic differences in socioeconomic and health determinants related to self-rated health status: A study on community-dwelling Israeli Jews and Arabs in old age. *International Journal of Environmental Research and Public Health, 19*(20). https://doi.org/10.3390/ijerph192013660

Salnikov, L., & Baramiya, M. G. (2021). From autonomy to integration, from integration to dynamically balanced integrated co-existence: Non-aging as the third stage of development. *Frontiers in Aging, 2*, 655315. https://doi.org/10.3389/fragi.2021.655315

Scarpina, F., D'Agata, F., Priano, L., & Mauro, A. (2021). Difference Between young and old adults' performance on the psychology experiment building language (PEBL) test battery: What is the role of familiarity with technology in cognitive performance? *Assessment, 28*(6), 1723–1734. https://doi.org/10.1177/1073191120918010

Sharma, N., Murari, G., Vandermorris, S., Verhoeff, N., Herrmann, N., Chen, J. J., & Mah, L. (2021). Functional connectivity between the posterior default mode network and parahippocampal gyrus is disrupted in older adults with subjective cognitive decline and correlates with subjective memory ability. *Journal of Alzheimer's Disease, 82*(1), 435–445. https://doi.org/10.3233/JAD-201579

Sheppard, C. L., Gould, S., Austen, A., & Hitzig, S. L. (2022). Perceptions of risk: Perspectives on crime and safety in public housing for older adults. *Gerontologist, 62*(6), 900–910. https://doi.org/10.1093/geront/gnab155

Sheshadri, A., & Elia, J. R. (2022). Physical resilience in older patients incident to hemodialysis: Can following trajectories improve our ability to intervene on functional decline? *Kidney International Reports, 7*(9), 1927–1929. https://doi.org/10.1016/j.ekir.2022.07.172

Siltanen, S., Tourunen, A., Saajanaho, M., Palmberg, L., Portegijs, E., & Rantanen, T. (2021). Psychological resilience and active aging among older people with mobility limitations. *European Journal of Ageing, 18*(1), 65–74. https://doi.org/10.1007/s10433-020-00569-4

Siqueira Junior, J. A., Lopes, A., Godtsfriedt, C. E. S., Justina, M. D. D., de Paiva, K. M., d'Orsi, E., & Rech, C. R. (2022). Neighbourhood walkability and mental health in older adults: A cross-sectional analysis from EpiFloripa Aging Study. *Frontiers in Aging, 3*, 915292. https://doi.org/10.3389/fragi.2022.915292

Stokols, D. (1992). Establishing and maintaining healthy environments. Toward a social ecology of health promotion. *The American Psychologist, 47*(1), 6–22. https://doi.org/10.1037//0003-066x.47.1.6

Taei, A., Jonson, H., & Granbom, M. (2022). Crime, disorder, and territorial stigmatization: Older adults living in deprived neighborhoods. *Gerontologist*. https://doi.org/10.1093/geront/gnac159

Turner, S. G., Jarrott, S. E., & Katz, B. (2023). Intergenerational programming increases solid food consumption for adult day center attendees. *Journal of Applied Gerontology, 42*(2), 160–169. https://doi.org/10.1177/07334648221134179

Verschuur, J., Li, S., Wolski, P., & Otto, F. E. L. (2021). Climate change as a driver of food insecurity in the 2007 Lesotho-South Africa drought. *Scientific Reports, 11*(1), 3852. https://doi.org/10.1038/s41598-021-83375-x

Vlachantoni, A., Falkingham, J., Evandrou, M., & Qin, M. (2023). Has COVID-19 been the great leveler? The changing use of intergenerational digital communications amongst older people in england during the pandemic. *Journal of Applied Gerontology*, 7334648231153385. https://doi.org/10.1177/07334648231153385
Ware, J. E. (1996). The SF-36 health survey. In B. A. Spilker (Ed.), *Quality of life and pharmacoeconomics in clinical trials* (2nd ed., pp. 337–345). Lipencott-Raven.
Wettstein, M., Wahl, H. W., & Siebert, J. S. (2020). 20-year trajectories of health in midlife and old age: Contrasting the impact of personality and attitudes toward own aging. *Psychology and Aging, 35*(6), 910–924. https://doi.org/10.1037/pag0000464
World Health Organization. (1998). *Programme on mental health: WHOQOL user manual.* Retrieved from Geneva: https://apps.who.int/iris/handle/10665/77932
World Health Organization. (2012). *The World Health Organization quality of life (WHOQOL) – BREF, 2012 revision.* Retrieved from Geneva: https://apps.who.int/iris/handle/10665/77773
Yang, T., Shi, H., Guo, Y., Jin, X., Liu, Y., Gao, Y., & Deng, J. (2019). Effect of work environment on presenteeism among aging American workers: The moderated mediating effect of sense of control. *International Journal of Environmental Research and Public Health, 17*(1). https://doi.org/10.3390/ijerph17010245
Zhu, C., Walsh, C., Zhou, L., & Zhang, X. (2023). Latent classification analysis of leisure activities and their impact on ADL, IADL and cognitive ability of older adults based on CLHLS (2008–2018). *International Journal of Environmental Research and Public Health, 20*(2). https://doi.org/10.3390/ijerph20021546

Chapter 4
Older People's Functionality and Community Participation: An Interdisciplinary Health-Transport Approach for Age-Friendly Cities

Antonia Echeverría, Paulina Del Solar, and Rodrigo Fernández

Acronyms and Abbreviations

COP25	25 Conference of the Parties to the United Nations Framework Convention on Climate Change
MTT	Ministry of Transport and Telecommunications
PRM	Persons with Reduced Mobility
PT	Public transportation
UK	United Kingdom
UN	United Nations
US	United States
WHO	World Health Organization

4.1 Introduction

The United Nations (UN) states that since 2007, more than half of the world's population in cities is expected to increase to 60% by 2030 (United Nations, n.d.). Accelerated urbanization, sometimes without long-term planning, has led to the appearance of inadequate infrastructure and services, affecting the living conditions

A. Echeverría (✉)
Facultad de Medicina, Escuela de Terapia Ocupacional, Universidad de Los Andes, Santiago, Chile

Millennium Institute for Care Research (MICARE), Santiago, Chile
e-mail: aecheverria@uandes.cl

P. Del Solar · R. Fernández
Facultad de Medicina, Escuela de Terapia Ocupacional, Universidad de Los Andes, Santiago, Chile
e-mail: pdelsolar@uandes.cl; rfa@uandes.cl

N. Asiamah et al. (eds.), *Sustainable Neighbourhoods for Ageing in Place*,
https://doi.org/10.1007/978-3-031-41594-4_4

of its inhabitants. In turn, the technological and industrial development of the globalization era has generated significant contributions to humanity, but at the same time, significant challenges due to the impact that this implies on the environment and people, especially older people. Catastrophes or global crises (understanding crisis as an unexpected event), increase the participation gap of old age, aggravated in many cases by the normative bio-physiological decline of this stage, which has repercussions on the capacity to adapt. Or, when faced with emergencies due to a catastrophe, mitigation actions and strategies do not consider the needs of the seniors, often leaving them excluded, which sometimes implies the denial of basic human rights (UN, 2020). These and other reasons have led the UN to declare the need to create sustainable, inclusive, safe, and resilient cities and communities as one of the 17 Sustainable Development Goals (UN, n.d.).

If functionality corresponds to the interaction of an older person with his environment (WHO, 2015), catastrophes or global crises are not alien to this segment of the population, which is increasing year by year. From this point, it is impossible to separate the older person's functionality from the context's elements. Hence, urbanization, industrialization, and modernization in which we live are phenomena that are not exempt from the analysis, considering that industrialization and innovation or modernization have begun to play a key role in creating sustainable cities and introducing new technologies for efficient use of resources. A clear example of this is the digital revolution that triggered the COVID-19 pandemic in a transversal manner in different sectors of society, such as health, education, essential goods, and services (UN, n.d.).

Evidence has shown that older people live better in their significant environments (Lewis & Buffle, 2020; Wiles et al., 2012), opening the challenge and opportunity to the States to plan, implement, and inhabit neighbourhoods, housing, and urban design that promotes the participation of older people in such environments in a meaningful and permanent way over time (UN, 1991). But to use the environment to promote aging in place, public transport (PT) is a prerequisite in many cases.

Since 2007, the World Health Organization (WHO) has declared the urgency of enabling "global age-friendly cities" by generating a guide to get localities to adapt their structures and services to "be accessible and inclusive of old age with diverse needs and capacities" (WHO, 2007, p. 6), being one of its thematic areas of transportation. During old age, PT takes a fundamental role in promoting participation. At some minute, older people with private vehicles lose their driver's licenses and/or willingness to use them (Edwards et al., 2008). Therefore, PT is the only way to participate in society (Tyler, 2015). Even more, evidence shows that lower use of PT is a relevant indicator of dependence (Edjolo et al., 2016). Since dependence is one of the most expensive and prevalent conditions in old age, it is worth asking how PT favours or limits its functionality, strengths and weaknesses, and challenges.

This chapter analyses the factors influencing community mobility and public transportation (PT) use, specifically among older people. It aims to explore PT's opportunities for participation in significant environments and examine its impact on the performance of instrumental daily life activities. The chapter also seeks to investigate the role of PT in designing and implementing age-friendly cities, as well as understand the perception of seniors regarding PT and how it affects their health condition, well-being, and overall quality of life. Furthermore, the chapter aims to provide insights into the recommendations proposed by international evidence and good practices to transform localities into more age-friendly and inclusive cities.

4.2 Public Transportation and Its Impact on the Participation of Older People

PT is vital in promoting health, active aging, social connectedness, and meaningful engagement, impacting well-being and independence. However, the decline in older adults' participation raises concerns about decreased environmental demands, potential loss of bodily functions, and increased social isolation.

In 2020, Ryvicker M. et al. estimated that approximately 2.3 million older people in the USA have barriers to using transportation, which could affect participation in social activities, visiting family or friends, attending religious services, or medical appointments (Luiu & Tight, 2021). This alerts given the vital role that PT plays in connecting people, especially those older people who abandoned private vehicle driving (Musselwhite, 2017). Thus, the use of PT is positioned as a vital component in health promotion given that it allows old age to remain active, socially connected, and involved in meaningful activities (Gajardo et al., 2012) experiences that impact health, well-being (Musselwhite, 2017), and independence (Luiu & Tight, 2021).

To the extent that older people stop participating in their significant environments, they perceive lower demands from the environment, enhancing the loss of bodily functions and structures due to disuse, which, in the stage of old age where per se a bio-physiological decline occurs, this situation becomes a relevant risk factor. The study by Gajardo et al. (2012) projects the impact of the non-use of PT on old age, mentioning that this could have repercussions on mood and self-esteem, reduce activities outside the home, and therefore favour social isolation, affecting the quality of life.

It is known that social isolation and loneliness in old age bring negative consequences for health and quality of life; therefore, favouring the maintenance of PT in the life course is a necessary action to facilitate social participation, especially in old age. For this, it is essential to make a detailed analysis of the links that involve mobility in the community to recognize the perceptions and difficulties presented by the seniors, with the ultimate goal of approaching the construction of sustainable cities that ensure the participation of all (Broome et al., 2010; Gajardo et al., 2012). In the following section, we define the contemporary way of understanding mobility in public transport through the so-called mobility chain.

4.3 Public Transport Mobility Perspective

From a transportation perspective, mobility is the ease of moving far and fast by any means of transportation, including PT, understood as railroads, subway systems, buses, cabs, shared cabs, and other nontraditional means, such as rickshaws (Tyler, 2015). Mobility in PT can be represented as a chain composed of several "links".

The first link is the accessibility to the PT system. It consists of the ease of reaching the PT system from the point of origin (e.g. the home) to the final destination (e.g. the urban activity centre). According to Tyler (2015), older people generally walk this journey, highlighting the importance of road and urban conditions to favour a safe and effective performance in this population segment. Hence, friendly cities must be adapted with urban elements such as wide and well-maintained sidewalks, regular pedestrian paths in green areas, and safe and properly marked crosswalks, among other aspects.

This stage of mobility is the most important for older people, because if there is no adequate infrastructure for them, there will be no option to use PT. And it is not only the infrastructure that matters but also its condition. A deteriorated sidewalk or a poorly located crosswalk constitutes significant limitations for an older person who presents, for example, problems with vision and balance or moves with some technical aid. Another essential condition for accessibility is walking distance. Tyler (2015) states that the maximum distance an older person using a stick can walk without taking a break is 200 m. If access to the PT system is more distant, resting places during the journey will be necessary.

Related to this, older people report that bus stops are located at long distances from their homes (Broome et al., 2010), an aspect that could be configured as a difficulty for walking (Gajardo et al., 2012), considering that in addition to the distance, the condition of sidewalks and crossings depends on the condition of the sidewalks and crossings to make this displacement safe. In this sense, intersectoral action at the state and local policy level becomes essential to ensure measures that consider the minimum necessary in road infrastructure and urban planning, to be age-friendly.

It should be noted that before accessibility begins, Gajardo et al. (2012) suggest that older people carry out a series of anticipatory actions to ensure their participation in the use of PT, such as knowing the route they will use to avoid improvisation. This situation could put them at risk. To this end, seniors seek access to information on the PT's routes, frequencies, schedules, and payment systems, which have advanced considerably with the modernization of the services. One element that the older people highlight from the modernization is incorporating an electronic payment system that allows them to charge an amount in advance and circulate with less money, thus reducing their exposure to violent situations such as robberies during their trips (Gajardo et al., 2012). But, on the other hand, for some people, it could be a barrier (Carney & Kandt, 2022) due to the need to access devices and the Internet in addition to being able to use them.

Seniors recognize they have more difficulties using the PT system during peak hours, so they temporarily modify its use, planning medical hours or other activities during off-peak hours. In this sense, an older person that uses the PT does not contribute to the overcrowding of vehicles but, on the contrary, uses this means when it is less in demand (Gajardo et al., 2012)

The second link of mobility is called vehicle access and means being able to enter and exit the PT system. It involves aspects such as waiting time for the service, identifying the service, and getting on and off the vehicles, such as cab stops, bus stops, and subway or train stations. The elements that contribute to this process are the existence of a place to wait for users, suitable areas for stopping vehicles, and information for the user at the stop, for example, that the sign is at an adequate height and the information contained is legible and clear. Stops can be part of the sidewalk or specially constructed locations. Is not only the presence of a bus stop significant but also its facilities, such as comfortable seating, protection from the weather (rain, sun, wind), lighting, and information that indicates the routes, schedules, and itineraries of the services. For seniors, content and legibility must be clear (typography, size and colour of the letters, contrast with the sign's background). In addition to the above, the height of the stop sign should be readable at the user's eye level, i.e. between 80 and 110 cm from the ground according to universal accessibility standards, a height that favours the reading of those who use wheelchairs (National Disability Authority, n.d.).

Regarding this link, older people perceive specific difficulties such as schedules and frequencies (Broome et al., 2010), waiting to stand at the bus stop (Luiu & Tight, 2021), long waiting times (Luiu & Tight, 2021), and crowds at the stop (Musselwhite, 2017). Given that older people plan their departures, alterations in schedules and/or low frequency of runs imply a greater demand for them in terms of altering the previous planning and/or extending the time of departure to carry it out. Those, as mentioned above, demand overexertion in physical, psychological, and/or emotional terms.

About access to the transport system, they highlight the difficulty that arises when entering or exiting the vehicles (Luiu & Tight, 2021) (Broome et al., 2010) (Musselwhite, 2017), especially when there are steps to enter or it stops far from the sidewalk (Musselwhite, 2017), since they are configured as a significant risk for falls. From this, it is essential to note the relevance of the planning and road infrastructure of a locality, since to the extent that using the PT means a risk for the seniors, at some point, the perception of risk in use will be greater than the benefit of moving, triggering the older people to stop participating or resort to an additional cost to use a private paid transport or ask for help and company to a third party. Situations such as these imply the social exclusion of older people in the use of public resources and community participation.

Older people also identified a need for bus stations/waiting stops to have adequate seating facilities, shade, and protection from weather conditions (Broome et al., 2010). As mentioned above, to the extent that the whole experience involves physical, psychological, and/or emotional overdemand, the motivation to perform the desired activity is diminished to the point of desisting from participating in social, family, and/or community activities because they do not feel able and/or safe to leave their homes.

The last link in the mobility chain is in-vehicle travel, the ease with which older people move smoothly in traffic once aboard PT vehicles. Contributing physical elements are comfortable PT vehicles, roads, and railways in good condition. An essential aspect of the comfort of old age is the quantity and quality of seats; for this, it is necessary not only to consider the seats reserved for people with reduced mobility, ranging from pregnant women and people with a temporary disability, but also to consider the seats related to the local reality and the proportion of existing older people, enabling as many seats as necessary to ensure a safe and comfortable trip. In the USA, bus service frequencies are designed so that 25% of the seats are always available for anyone (Vuchic, 2007). In general terms, seniors are more prone than younger ones to balance and gait disturbances, taking into account, in addition, that they have slower protective reactions, as being able to be seated is a priority when traveling in PT (Karekla & Tyler, 2018).

Regarding this link, older people highlight stories oriented to the discomfort and insecurity regarding their infrastructure, given that there were structural aspects that did not meet their needs with descriptions, such as that strollers or objects sometimes obstructed the displacement inside the vehicle, or there was such a crowd that did not allow a smooth and safe mobilization to the priority seats (Musselwhite, 2017).

Along the same lines, the attitudes of the social environment are also an important element to consider. The attitude of the drivers was highlighted (Broome et al., 2010), given that it determines, for example, the time they wait to start the machine once the older person has entered it (Musselwhite, 2017) and whether or not they wait to be seated. And on the other hand, the other passengers stand out as a fundamental factor in facilitating performance and getting a seat (Musselwhite, 2017).

In short, if any of these links in the mobility chain is deficient, the PT system will be highly complex and/or nonexistent for older people. We hypothesize that old-age mobility in friendly cities can be improved through small changes applied to each link in the mobility chain. In the first instance, it is optional to replace the entire bus fleet; it would be sufficient to improve the bus stops that connect areas where older people live with community centres, for example.

By addressing the specific needs at each link of the mobility chain, significant improvements can be made to enhance the overall mobility experience for older individuals, promoting their well-being and social inclusion.

4.4 Public Transport Routes and Their Impact on the Functionality of Older Persons

Another aspect that impacts the mobility of older people is the tendering of PT routes, related to the connectivity involved in the layout of the mobility trajectory in the city. This situation requires a design capable of integrating the main community participation environments with the road infrastructure of the place. Two approaches can be observed in this regard. The traditional approach maintains the nineteenth-century practice of railroad tracks, as is still done today with bus routes and subway lines. The aim is to connect terminals by a route that is as straight as possible and with stops as far apart as possible to ensure high commercial speed (average travel speed between two points, including any intermediate delays). In this approach, accessibility to activity centres, such as commerce, public services, and health centres, must be done by walking to and from the nearest stop. In this regard, the Center for Sustainable Urban Development indicates that only 5.6% of older people in Santiago de Chile have a metro station within a 10-min walk. This percentage drops to 3.6% in the case of low-income older people. Another source points out that in Santiago, more than 93% of the older people are within 1000 m of a bus stop or metro station National Socioeconomic Characterization Survey (Encuesta de Caracterización Socioeconómica Nacional de Chile, [CASEN]) (MIDESO, 2018). Assuming a senior walking speed of 0.9 m per second, according to the National Traffic Safety Commission (Comisión Nacional de Seguridad de Tránsito [CONASET], 2013), this means almost 20 min of walking.

On the other hand, the contemporary approach for tendering PT routes is to connect intermediate demand points. In this case, the objective is to provide better accessibility to these points, which, in turn, define the location of the main stops. The rest of the stops are located along the route at walking distance to the demand places. In this regard, the UK policy states that every London house should be 400 m from a bus route (Transport for London, 2017). However, on a grid network of streets, the distance to the nearest bus stop is 566 m (Tyler, 2015). This design principle has gradually been adopted in the European Union and extended to the light rail systems. In the case of the USA, on the other hand, the standards refer only to the distance between PT stops. Thus, for buses, the spacing of bus stops is set between 300 and 500 m and between 500 and 2000 m between subway stations (Vuchic, 2007). However, these standards do not mention how far the origin or destination of the trip should be from the route.

Older people describe the wrong location of bus stops and routes, resulting in the need for more direct services to specific areas such as hospitals, commerce, and the city centre (Broome et al., 2010). Because of this, to reach their destination, they had to walk yet another distance. In summary, the tendering of PT routes has a significant impact on the mobility of older people, particularly concerning connectivity and the layout of the transportation network in cities.

4.5 The Case of Chile

This section describes how policies for including older people in PT have been carried out in Chile as an example of a Latin American country, particularly in the capital city of Santiago. Chile has great geographical and population diversity due to its territorial extension (the longest country in the world) and climatic conditions. The following section will analyse the PT system of Santiago, the country's capital.

The current PT system of Santiago consists of a subway system of 7 lines, 142 km, and 138 stations, 99% of them with elevators to get from the street to the platform (Directorio de Transporte Público Metropolitano [DTPM], n.d.). It is complemented by a bus system with a fleet of 6900 vehicles, of which 1800 are electric, covering almost 3000 km of streets and 11,000 bus stops (DTPM, n.d.). Nearly 90% of the buses are low-floor, without steps, and with seats and priority areas for persons with reduced mobility (PRM), among which older people have been included (DTPM, 2020).

The document that has guided the design of PT in Chile is the National Transportation Policy (Política National de Transporte, [PNT]) of the Chilean Ministry of Transport and Telecommunications (MTT, 2014). This policy does not explicitly mention older people but refers to PRM. However, seniors can be included within PRM, using the criteria of the National Service for Disability (Servicio Nacional de la Discapacidad [SENADIS], 2017), which defines PRM as those who "have, for any reason, difficulty in moving, either temporarily or permanently. For an approach to its quantification, reference will be made to three identifiable population groups: older people (i.e., people over 60 years old), women in an advanced state of gestation, and families with children up to 3 years old and large people".

According to the above, one of the proposals of the PNT is to ensure the mobility of PRM, indicating that the design of public transport systems should include facilities for use by people with reduced mobility, guaranteeing the correct use of the regulations in the design of infrastructure and information, incorporating adapted vehicles, and defining services consistent with special needs' (MTT, 2014, p. 34). Regarding pedestrian accessibility, the primary way to achieve PT, the PNT indicates that pedestrians should include special plans for people with reduced mobility, allowing them to use pedestrian facilities and connections with PT (MTT, 2014).

The National Strategy for Sustainable Mobility (Estrategia Nacional de Movilidad Sostenible [ENMS]) (MTT, 2021) is a complementary instrument to other existing strategies and policies in Chile. It is aligned with other documents, such as the report of the COP25 Cities Roundtable. One of the proposals of the document is to have a universally accessible PT, which considers the updating of standards for vehicles and the modernization of fleets to move towards a universally accessible PT. In another aspect, it is thought that infrastructure and public spaces should be universally accessible to ensure compliance with the standards of transportation infrastructure and public spaces for proper use by users with reduced mobility and adapting existing infrastructure to users with disabilities. And finally, the need to have a safe and inclusive PT is mentioned. A PT system that allows its safe use without distinction of users. Abolishing violence, whether gender-based or against minorities, such as the elderly in PT-related spaces, such as vehicles, access roads, stops, and stations. Improving these elements seeks to minimize the exposure of users to risk, harassment, or violence (MTT, 2021).

Bearing in mind that accessibility is the link in the mobility chain that is made by walking, a seminal study on pedestrian accessibility in Chile (CITRA Ltda., 1999) took a sample of pedestrians, 5% of whom were over 65 years old, to quantify the valuation of various attributes of walking, such as environment, noise, lighting, surveillance, crosswalks, width and quality of sidewalks, and pedestrian congestion. One of the findings was that people over 60, i.e. choose their pedestrian route primarily based on the fewer streets that need to be crossed to reach their destination. Regarding accessibility to PT, this destination is the bus stop or subway station.

Regarding PT access and circulation, Fernandez et al. (2022) address priority measures for the expeditious circulation of buses in developing countries, such as exclusive lanes for bus circulation, adjusting traffic light timings to coincide with the arrival of the bus at the crossing and the design of accessible bus stops. Some of their recommendations have been adopted in Chile, favouring older people by extension; for example, elevated bus stop places for waiting with curved curbs so that buses stop at the side to make it easier for passengers to board and alight (Fernandez et al., 2022). However, most bus stops need old age-friendly conditions, such as covered shelters, seating, and understandable signage. Only a few bus stops have waiting areas at the same level as the vehicle. But at the rest of the bus stops, the buses stop far from the sidewalk, and, consequently, passengers must get off the sidewalk onto the street and, from there, get on the bus. Getting off is even more uncomfortable as it involves stepping down from a height of 25 cm between the bus and the street. If this is uncomfortable for anyone, it is even more so for older people (Fernández et al. (2022).

In a mixed design study, Vecchio et al. (2020) rescued perceptions of a group of older people related to their experience with the built environment during mobility in Santiago, rescuing that this often does not allow safe displacements. The poor condition of the sidewalks configures a scenario of imminent risk of falls for seniors and even more in those with restricted mobility (Vecchio et al., 2020), a perception that is congruent with what has been exposed at the international level. The social environment in this link also takes a significant position, referring to the fact that in peak traffic hours, people circulate at higher speeds feeling vulnerable to falls, in addition to preferring to go out during daylight hours to avoid exposing themselves to situations of violence such as an assault (Public Innovation Lab, 2018), and that coexistence with other modes of transport, such as bicycles, scooters, skateboards, among others, is also complex at times (Vecchio et al., 2020).

According to the CASEN Survey conducted in 2017, 98.3% of households in Chile are located at a distance of fewer than eight blocks or 1 km from a PT service; however, older people perceive difficulties in accessing it (MIDESO, 2018).

The barriers that stand out in the access link in Chile are similar to the one described above, with older people referring that crowds at stops are an element that hinders their performance (Gajardo et al., 2012), getting on and off vehicles given the presence of steps (Gajardo et al., 2012), as well as the obstacle presented when buses stop distant from the sidewalk (Laboratorio de Innovación Pública, 2018). A striking action described by Gajardo et al. (2012) was that some older people sought to get off at the front so that the driver could see them and respect their descent times to avoid accidents.

Related to the physical aspects of the vehicle, alluding to the mobility circulation link, Gajardo et al. (2012) reveal that the characteristics of the seats, their inclination, nonslip characteristics, their countermarch position, as well as the height of the handrails, were not optimal to provide safety in the use of the buses. On the other hand, regarding the displacement of the PT, some older people highlight as a problem the poor cushioning of the buses (Gajardo et al., 2012), since this generates sudden and surprising movements inside the bus, making it unfriendly for them.

In conclusion, the PT system in Santiago, Chile, has improved accessibility for persons with reduced mobility, including older people. The National Transportation Policy and Strategy for Sustainable Mobility provide guidelines and proposals to ensure PT users' mobility, safety, and inclusivity. However, further improvements are necessary to create a universally accessible and age-friendly PT system in Chile.

4.6 Mobility and Its Link with the Recommendations of Age-Friendly Cities

According to what has been reviewed, even when the PT has been modernized and improvements have been incorporated into its system, seniors continue perceiving barriers to its use, impacting the performance of their meaningful activities and quality of life. In this section, using the same mobility links explained above and gathering the perceptions/experiences of the older people in the PT, a connection is made with the WHO recommendations regarding age-friendly cities.

Regarding the first link of mobility, everything related to "accessibility", in the Global Age-Friendly Cities Guide, two relevant points are raised: access to stops/stations and their location in the layout. Regarding the "stops" and PT stations, it is highlighted that the location of these should be close to the older people's homes as well as the access infrastructure, such as sidewalks and crossings, should be in good condition to prevent this from becoming a barrier to the performance of this occupation (Gajardo et al., 2012). As a second aspect, the routes should cover the most frequent or relevant travel destinations, given that, as mentioned above, the traditional approach makes a straight route that does not necessarily include the priority participation points for the seniors. Therefore, it is necessary to have older people in the design of the routes and to promote the creation of routes under the contemporary approach, which identifies and links those places with intermediate demand and thus promotes maximum social and community participation.

Another relevant point for this first mobility link is all the actions older people perform before leaving home. The guide mentions that "information" is a cardinal aspect, i.e. older people have information on transportation systems, how to use them, and their schedules. Currently, the trend is to digitize this information, so a gap may arise between those older people that do not use the Internet and/or need better connectivity to it and those that do. To subsidize this, Broome et al. (2010) suggest the need to incorporate "education" to older people regarding the transportation system, and today, it would also be necessary to integrate education regarding the use of technology in it to access updated information regarding route planning, operating schedules, and waiting times for the arrival of transportation.

Now, addressing the access link in mobility, the guide highlights as necessary the "reliability" (certainty that the PT will pass) and "frequency" (time ranges of the PT or traffic flow through the stop), which, as previously exposed, represents a difficulty for older people when this fails. The WHO states that "the availability of public transport services should be frequent and reliable" to avoid crowds and long waiting times. In this sense, each country/region/commune should maintain systems of continuous monitoring of transport vehicle frequencies and have a response mechanism in the face of traffic contingencies and information regarding the arrival of the PT vehicle.

Regarding stops and stations, it is essential to consider the design and condition of the stops to ensure a smooth use; for example, they protect against weather conditions or have seats for resting while waiting for the PT vehicle in case of traffic jams due to long waits, so that the seniors can find a safe and comfortable place to stay, without the exit triggering a physical, cognitive, or emotional overdemand that adds stress to the unforeseen events that may occur.

A difficulty reported by older people for access to the PT is the ascent/descent of the vehicle. This aspect is addressed by the WHO recommendations both related to the "behaviour of drivers" to place the vehicle close to the curb, as well as related to the presence of steps in the vehicles themselves, which make the task of entering or exiting complex, so they should consider that their infrastructure is "friendly to the old age".

For the last link in the mobility chain, referred to as "circulation", the guide reinforces the need for vehicles to have preferential seating for older people so that they can travel safely and comfortably. These facilities should be accompanied by education of the general population regarding old age, so that they can understand the difficulties that this age group may present in mobility and thus understand the need for using a preferential seat and that these are respected. In the same vein, drivers of vehicles should also be specially trained about older people, for example, to ensure that they stop the vehicle at the correct distance from the curb and wait a prudent time for the vehicle to start (until the older person has taken a seat).

Despite the modernization and improvements in the PT system, older people still face barriers that impact their ability to use PT effectively, affecting their quality of life. The WHO recommendations for age-friendly cities highlight the importance of accessibility, information, reliability, and comfort in PT design. Addressing these aspects, it is crucial to ensure that stops and stations are conveniently located, routes cover relevant destinations for older people, and information is accessible. The reliability and frequency of PT services should be maintained and stops should be designed to provide shelter and seating. Vehicle accessibility, including easy boarding and preferential seating, is essential, accompanied by education for the public and drivers to understand and support older people's needs. By addressing these factors, PT can become more age-friendly and enhance the mobility experience for older people.

4.7 Discussion

Crisis or emergencies, regardless of their natural origin (such as climatic phenomena) or human head (such as wars), strain those services of public use within a community.

The most representative case at the world level today could be the COVID-19 pandemic. The quarantine policies and physical and social distancing implemented challenged authorities around the world to reorganize services, one of the most conflictive during the entire period of the pandemic and even to this day, PT. During the pandemic, airports were closed; work and studies began to be done remotely and purchases of all kinds were made virtually, among many other examples. This abrupt and unexpected change affected the entire world: from early childhood to older people.

During the pandemic, restrictive measures to prevent infection, such as quarantines and physical and social distancing policies, were especially explicit and highly demanding for healthcare institutions and public transport. A study from Sweden showed a 40–60% drop in the number of PT users in all regions of the country (Böcker et al., 2023). In cities in the United Kingdom, it was reported that 2 years after the onset of COVID-19, the number of PT users was still below pre-pandemic use (Long et al., 2023). In this scenario, the pandemic prompted individuals to seek other modes of transportation, such as walking, bicycling, or car use, as people sought to ensure their health, safety, and comfort (Gao et al., 2023). In the case of New York, a study revealed that the major transportation concerns were the following: mask use and social distancing, virus/infection transmission, and health concerns when using the bus (Gao et al., 2023). According to Böcker et al. (2023) in the use of PT, older people were most affected by the pandemic and its restrictions, positioning them as a highly vulnerable group. When associating the above with health, the situation is further aggravated, because, to the extent that contact with the environment is eliminated, functionality is affected, increasing the risk of loss of intrinsic capacity due to disuse. This leads to functional deterioration and the appearance of dependence in old age.

In the case of PT during the pandemic, in addition to care measures such as the use of masks and physical distancing, an aspect practically impossible to control in PT, seats were blocked to avoid contagion. These conditions increased the population's perception of insecurity when using PT. According to the study by Böcker et al. (2023), in Seoul (Korea), the senior citizens were the ones who most limited their use of the subway during the pandemic period. In the case of Canada, it was reported that older people who were able to resume the use of private transport stopped using public transport as soon as the pandemic was declared. This is a reality that cannot be generalized to all.

In measuring the perception of safety (at the station, onboard, and of suffering accidents), the study by Böcker et al. (2023) shows that the old-age population has comparatively higher levels of insecurity than the other age groups, higher ranks before and during the pandemic at all three levels (with an increase in the pandemic).

The study by Long et al. (2023) in the city of Birmingham (UK), where access to buses is free for older people, studied the behaviour of use according to age, gender, and ethnicity, finding that there are significant differences. The pre-pandemic profile found was as follows: women, between 71 and 75 years old, white. After analysis, it was found that, comparatively, women stopped using transport more than men, older people (>85 years) than those aged 66–70 years, and white people more than other ethnicities. More than half of those >70 years old (men and women) avoided PT during the pandemic and did not return. Similar results were found by Gao et al. (2023) in the USA regarding a decrease in public transport use by older people during the pandemic. They highlighted that, during the pandemic, older people recorded a comparatively higher number of trips than the rest of the adults, a situation that could be explained by the digital divide in accessing basic services such as shopping for merchandise, pharmacy, and medical visits, among others. Another interesting finding was the increase in the use of bicycles. Although the increase in use by older people was not as significant as in younger adults, the increase in use could be related to health and economic limitations to using other available means of transportation. The study by Shaer and Haghshenas (2021) describes that older people used bicycles as a means of getting around on long trips without having to expose themselves to public transport measures or insecurity of contagion. The bicycle proved to be a resilient alternative to public transport problems and the impossibility of using private transport. In addition, it is an environmentally friendly means of transport.

The studies, regardless of the aspects investigated and countries, coincide in that the older people made changes in their mobility pattern: changes in routes, means of transportation, and schedules of use, among others. Measures that sought to reduce the chances of contagion and enhance the perception of the safety of care during transportation (Böcker et al., 2023; Gao et al., 2023; Long et al., 2023).

The COVID-19 pandemic, which affected the entire world, brought to light sociodemographic and territorial inequalities in one of the most important instrumental activities of daily living for old age, which is the use of transportation. A longitudinal study following a cohort for 24 years on more than 3000 older people in France, stated among other activities that losing the ability to make use of transportation, constitutes the gateway to dependence (Edjolo et al., 2016). From this, the challenges for countries at the intersectoral level are varied, starting from the fact that feeling safe is a human right and part of the Sustainable Development Goals; therefore, it will be the duty of the authorities to plan a PT system that allows older people to feel safe and thus maintain mobility in the community and thanks to this, participation in activities of interest in their significant environments. But also, how policymakers in urban planning have to consider the socio-demographic and territorial diversity when designing age-friendly cities, combining also the accessibility to services in safe routes that favour the displacement and daily performance of older people in the face of future crises.

4.8 Conclusion

Within the framework of the international challenges regarding sustainability, Chile has launched its National Strategy for Sustainable Mobility, which states the objectives and measures to be applied in the country to favour environmental, social, and economic integration, which in its interactions demands an implementation that is viable, bearable, and equitable over time, since only in this way will sustainability be achieved. Considering that Chile is a country where access to goods and services is increasing, PT is not only the most affordable and most used means of transportation by low-income people, but it is also highly cost-effective in terms of pollution, since it is the one with the lowest amount of emissions per passenger related to kilometres travelled. However, it is less used than others (an average of 30% of total trips at a national level, that is, only 1 out of every three trips are made in PT according to MTT). When the conditions of use are lower than the conditions of sustainability, there is a risk that decision-makers do not prioritize the changes required to favour an efficient, safe, and quality PT for the inhabitants. According to what we have been able to analyse in the development of this chapter, this is especially relevant in older people who adopt this means as the highest priority within the system. The reasons are accessing (it has a greater number of access points than other means), cost (which is generally lower than other means), and habituation (habit of use concerning other means of transport). However, with age, safety in use begins to play a dominant role in the daily performance of older people and often determines the exploration of private transport.

Returning to the definition of PT, as the one that allows connecting people with places where they need to perform an activity, the loss of this travel space constitutes a potential risk of functional deterioration since the fact of not being able to use PT will reduce the opportunities for participation and performance of seniors, restricting opportunities to maintain body functions and structures, skills, as well as significant roles that slow down the average physiological decline of aging. This situation would affect the five principles of the human rights of older persons. Given the relevance of PT for older people and the impact its disruption generates on older people's quality of life, well-being, and functional capacity, governments should formulate anticipatory coping strategies for crisis or catastrophic situations, such as the recent COVID-19 pandemic, to preserve as much as possible the use of public transport in old age. Considering the elderly in the design of remedial actions in the face of a world in constant crisis will favour the participation of the community with all its social diversity, in sustainable neighbourhoods avoiding their exclusion and segregation.

In this regard, States are responsible for integrating the intersectoral visions that favour and, in turn, impact the PT system even more when considering the segments of the population that make the most important use of it. Although the actions to be implemented to ensure age-friendly transportation are neither low-cost nor few, about what is proposed by international organizations as good practices in this area, the return can be quantified in terms of the health of the older people, since by maintaining participation in meaningful activities and roles, functionality is maintained and the onset of dependence of an increasingly growing segment is slowed down.

Review Questions

1. Explain the three links of the mobility chain in public transport and its relationship with older people.
2. Observe a bus stop that you know or are accustomed to using, and evaluate what opportunities for improvement you detect in older people based on the elements reviewed in this chapter.
3. Perform the analysis of a journey from the starting point (e.g. your home) to a bus stop or vice versa. Identify barriers and strengths related to the contents seen in the chapter.
4. In the means of transport you usually use, what are the elements that facilitate/restrict the participation of older people?
5. Reflect on the impact it would have on you, as a future older person, to stop using public transport: what would your routine be like if you could not use public transportation? To what extent would it impact you to stop doing these meaningful activities? What repercussions would it have on your health, well-being, and quality of life to stop doing these activities that are important and/or satisfying for you?

References

Böcker, L., Olsson, L. E., Priya Uteng, T., & Friman, M. (2023). Pandemic impacts on public transport safety and stress perceptions in Nordic cities. *Transportation Research. Part D, Transport and Environment, 114*, 103562. https://doi.org/10.1016/j.trd.2022.103562

Broome, K., Nalder, E., Worrall, L., & Boldy, D. (2010). Age-friendly buses? A comparison of reported barriers and facilitators to bus use for younger and older adults. *Australasian Journal on Ageing, 29*(1), 33–38. https://doi.org/10.1111/j.1741-6612.2009.00382.x

Carney, F., & Kandt, J. (2022). Health, out-of-home activities and digital inclusion in later life: Implications for emerging mobility services. *Journal of Transport & Health, 24*, 101311. https://doi.org/10.1016/j.jth.2021.101311

CITRA Ltda. Consultores en Ingeniería de Transporte Ltda. (1999). *Análisis y proposición de políticas de inversión para favorecer a los peatones.* http://www.sectra.gob.cl/biblioteca/detalle1.asp?mfn=694

Comisión Nacional de Seguridad de Tránsito (CONASET). (2013). *Manual de Señalización de Tránsito*. https://www.mtt.gob.cl/wp-content/uploads/2020/04/Manual-de-Sen%CC%83alizacion-de-Transito.pdf

Directorio de Transporte Público Metropolitano (DTPM). (2020). *Informe de Gestión 2020*. https://www.dtpm.cl/descargas/memoria/informe%20de%20gestion%202020.pdf

Directorio de Transporte Público Metropolitano (DTPM). (n.d.). *Contenido Esencial-Vías*. Retrieved June 6, 2023, from https://www.dtpm.cl/index.php/113-servicios

Edjolo, A., Proust-Lima, C., Delva, F., Datigues, J. F., & Pérès, K. (2016). Natural history of dependency in the elderly: A 24-year population-based study using a longitudinal item response theory model. *American Journal of Epidemiology, 183*(4), 277–285.

Edwards, J. D., Ross, L. A., Ackerman, M. L., Small, B. J., Ball, K. K., Bradley, S., & Dodson, J. E. (2008). Longitudinal predictors of driving cessation among older adults from the ACTIVE clinical trial. *The Journals of Gerontology Series B: Psychological Sciences and Social Sciences, 63*(1), P6–P12. https://doi.org/10.1093/geronb/63.1.P6

Fernández, R., Seriani, S., & Valencia, A. (2022). *Recomendaciones de Buenas Prácticas para Prioridades a Buses en Países en Desarrollo*. Nota Técnica, Sociedad Chilena de Ingeniería de Transporte (Sochitran).

Gajardo, J., Navarrete, E., López, C., Rodríguez, J., Rojas, A., Troncoso, S., & Rojas, A. (2012). Percepciones de personas mayores sobre su desempeño en el uso de transporte público en Santiago de Chile. *Revista Chilena de Terapia Ocupacional, 12*(1), Article 1. https://doi.org/10.5354/0719-5346.2012.22055

Gao, J., Lee, C. D., Ozbay, K., Zuo, F., & Chippendale, T. L. (2023). Understanding the travel challenges and gaps for older adults during the COVID-19 outbreak: Insights from the New York City area. *Transportation Research Interdisciplinary Perspectives, 19*, 100815. https://doi.org/10.1016/j.trip.2023.100815

Karekla, X., & Tyler, N. (2018). Reducing non-collision injuries aboard buses: Passenger balance whilst walking on the lower deck. *Safety Science, 105*, 128–133. https://doi.org/10.1016/j.ssci.2018.01.021

Laboratorio de Innovación Pública. (2018). *RedActiva: Fomentando la movilidad de los adultos mayores en la ciudad*. https://politicaspublicas.uc.cl/content/uploads/2018/03/RedActiva-2.pdf

Lewis, C., & Buffel, T. (2020). Aging in place and the places of aging: A longitudinal study. *Journal of Aging Studies, 54*, 100870. https://doi.org/10.1016/j.jaging.2020.100870

Long, A., Carney, F., & Kandt, J. (2023). Who is returning to public transport for non-work trips after COVID-19? Evidence from older citizens' smart cards in the UK's second-largest city region. *Journal of Transport Geography, 107*, 103529. https://doi.org/10.1016/j.jtrangeo.2023.103529

Luiu, C., & Tight, M. (2021). Travel difficulties and barriers during later life: Evidence from the National Travel Survey in England. *Journal of Transport Geography, 91*. Scopus. https://doi.org/10.1016/j.jtrangeo.2021.102973

Ministerio de Desarrollo Social (MIDESO). (2018). *Síntesis de resultados CASEN 2017: Vivienda y entorno* (p. 110). http://observatorio.ministeriodesarrollosocial.gob.cl/storage/docs/casen/2017/Resultados_vivienda_casen_2017.pdf

Ministerio de Transportes y Telecomunicaciones (MTT). (2014). *Política Nacional de Transportes: Conectando Chile*. https://www.mtt.gob.cl/wp-content/uploads/2013/05/documento-politica.pdf

Ministerio de Transportes y Telecomunicaciones (MTT). (2021). *Estrategia Nacional de Movilidad Sostenible (ENMS)*. https://www.subtrans.gob.cl/wp-content/uploads/2021/12/documento-ENMS-2-1.pdf

Musselwhite, C. (2017). Public and community transport. *Transport and Sustainability, 10*, 117–128. Scopus. https://doi.org/10.1108/S2044-994120170000010004

National Disabilities Authority. (n.d.). *1.2 Ensure that displays are within sight of people of all heights and people sitting in a wheelchair or buggy* | Centre for Excellence in Universal Design. Retrieved June 6, 2023, from https://www.universaldesign.ie/technology-ict/archive-irish-national-it-accessibility-guidelines/Telecoms/Guidelines-for-Telecoms-Accessibility-Priority-1/1-2-Ensure-that-displays-are-within-sight-of-people-of-all-heights-and-people-sitting-in-a-wheelchair-or-buggy/1-2-Ensure-that-displays-are-within-sight-of-people-of-all-heights-and-people-sitting-in-a-wheelchair-or-buggy.html

Ryvicker, M., Bollens-Lund, E., & Ornstein, K. A. (2020). Driving status and transportation disadvantage among medicare beneficiaries. *Journal of Applied Gerontology, 39*(9), 935–943. https://doi.org/10.1177/0733464818806834

Servicio Nacional de la Discapacidad (SENADIS). (2017). *Plan Chile Accessible: Bases metodológicas para la gestión de un plan territorial de accesibilidad.* https://www.senadis.gob.cl/descarga/i/4722/documento

Shaer, A., & Haghshenas, H. (2021). The impacts of COVID-19 on older adults' active transportation mode usage in Isfahan, Iran. *Journal of Transport & Health, 23*, 101244. https://doi.org/10.1016/j.jth.2021.101244

Transport for London (TfL). (2017). *Accessible bus stop design guidance.* https://content.tfl.gov.uk/bus-stop-design-guidance.pdf

Tyler, N. (2015). Buses and the city. In *Accessibility and the bus system: Transforming the world* (Vol. 1, pp. 3–45). ICE Publishing. https://doi.org/10.1680/aabs2ed.59818.003

United Nations (UN). (2020). *Decade of healthy ageing 2020–2030.* https://cdn.who.int/media/docs/default-source/decade-of-healthy-ageing/decade-proposal-final-apr2020rev-es.pdf?sfvrsn=b4b75ebc_25&download=true

United Nations (UN). (n.d.). *Goal 11- Sustainable cities and communities.* Sustainable Development Goals. Retrieved February 19, 2023, from https://www.un.org/sustainabledevelopment/es/cities/#tab-8dd6cb9078e4c78159c

United Nations Principles for Older Persons (Resolution 46/91). (1991). https://www.ohchr.org/sites/default/files/olderpersons.pdf

Vecchio, G., Castillo, B., & Steiniger, S. (2020). Movilidad urbana y personas mayores en Santiago de Chile: El valor de integrar métodos de análisis, un estudio en el barrio San Eugenio. *Revista de Urbanismo, 43*, Article 43. https://doi.org/10.5354/0717-5051.2020.57090

Vuchic, V. (2007). Transit system performance: Capacity, productivity, efficiency, and utilization. In *Urban transit systems and technology* (pp. 149–201). https://doi.org/10.1002/9780470168066.ch4

Wiles, J. L., Leibing, A., Guberman, N., Reeve, J., & Allen, R. E. S. (2012). The meaning of "aging in place" to older people. *The Gerontologist, 52*(3), 357–366. https://doi.org/10.1093/geront/gnr098

World Health Organization (WHO). (2007). *Global age-friendly cities: A guide* (p. 73). https://apps.who.int/iris/handle/10665/43805

World Health Organization (WHO). (2015). *World report on ageing and health.* https://apps.who.int/iris/bitstream/handle/10665/186466/9789240694873_spa.pdf

Chapter 5
Two Sides of the Coin in Ageing in Place: Neighbourhood Safety and Elder Abuse

Mehmet Öçal and Özge Kutlu

Abbreviations

EU	European Union
EUROSTAT	European Community Statistical Office
EU-SILC	European Union Statistics on Income and Living Conditions
ILCS	Income and Living Conditions Survey
TSI	Turkish Statistical Institute
WHO	World Health Organization

5.1 Introduction

What is the ultimate premise of the humanitarian mission and vision of today's states? The answer to this short but striking question is so simple: *welfare*. Each state, committing to social policy's humanitarian pursuit, targets to raise its citizens' welfare. So, how should welfare be defined in today's world? Indeed, it is a concept difficult to define since it does not rely on only one parameter. It may not be prudent to propose that a high-income person certainly has high welfare. Or, a healthy person cannot be regarded as one with high welfare based on the health parameter alone. Welfare refers to human values and to what extent one lives a decent life, that is, the "life they desire" by being protected from social risks. As a duty of the state,

M. Öçal
Department of Social Work, Burdur Mehmet Akif Ersoy University, Burdur, Turkey
e-mail: mocal@mehmetakif.edu.tr

Ö. Kutlu (✉)
Elderly Care Program, Burdur Mehmet Akif Ersoy University, Burdur, Turkey
e-mail: okutlu@mehmetakif.edu.tr

N. Asiamah et al. (eds.), *Sustainable Neighbourhoods for Ageing in Place*,
https://doi.org/10.1007/978-3-031-41594-4_5

it may remain an insufficient action to protect welfare. While the "social state" undertakes the mentioned duty, improving one's welfare can be shown as an act of the "welfare state". Yet, whose welfare do States guarantee with social policy? Despite changing by welfare regime typology, social policy's humanitarian pursuit is now constructed with a universal motive leaving all variables aside beyond the understanding of "citizenship". Social policy guarantees a minimum level of welfare for all, but what about those who have welfare relatively far below society?

Unfortunately, not everyone has the same level of welfare in society, but social policy sets a minimum level of welfare for everyone. If one goes below that minimum level of welfare, social policy is deemed to assist in improving it. More comprehensive social policies come into play for vulnerable groups more likely to confront more social risks than the general population, such as the disabled, the unemployed, ex-convicts, immigrants, children, and older adults. Ageing is an entropic and inevitable social risk, and people who are weakened by ageing are likely to be deprived of sufficient income, lose their ability to work, fail to protect themselves, and ultimately become in need of care. Increased needs with ageing and the corresponding decrease in income and strength may explain why one may need health and social care. Ageing also contributes to the risk of being abused. Thus, housing health and neighbourhood safety become essential for older adults spending most of their time at home and mandate the implementation of tailored policies. In this sense, this chapter aims to reveal the housing and neighbourhood welfare of older adults in the European Union (EU) and Türkiye and propose social policies, credited by "exemplary practices" of the EU, to prevent elder abuse and improve their neighbourhood safety in Türkiye.

The chapter aims to bring an approach to the issue of elder safety and abuse based on the environment in which the elderly person lives. In order to ensure ageing in place in a manner befitting human dignity, "neighbourhood safety parameters" classification has been made for elderly individuals. This classification has been designed by bringing together the elements of "housing health and safety, health and safety of the residential environment and older adults' socioeconomic status", which affect the neighbourhood safety of older adults. In line with the statistical measurements of each parameter, the steps, practices, and policies that should be taken for a decent ageing environment are discussed in the context of Türkiye.

This chapter comprises six sections. The first section sets the background where the rationale and aim of the chapter are presented. The second section then presents the focus and theoretical framework of the chapter. The methodology and findings are presented in the third and fourth sections, respectively, whereas a discussion of the findings is performed in the fifth section. Concluding remarks are then presented in the sixth section.

5.2 Study Focus and Theoretical Approach

In the twenty-first century, the world is literally experiencing a demographic transformation called global ageing, leading to *greying of the world population*. Although different parts of the world experience ageing at different levels, it is now undeniable that the world is ageing. Despite some opportunities, an increased older adult population cause the physical, psychological, and social problems of old age to become visible. Problems specific to ageing and older adults affect not only developing countries but also developed countries. Yet, developed countries seem to cope with such problems relatively more easily than developing countries, thanks to their robust human and economic resources. However, developing countries, such as Türkiye, confront ageing-specific issues more dramatically due to underdeveloped social security and support systems, narrowing family structure, and reduced family support (Keskinoğlu et al., 2004).

On the other hand, the grey tsunami of the world brings with it an increase in demand for some service areas, particularly healthcare services. Grey tsunami is emerging as a concept that will affect the whole world. The concept of grey tsunami highlights the problems that may occur in the future related to the ageing population (Öztuna, 2017). One of these problems is the increasing demand for different services. Finding a balance between elevated service needs and reduced premium inputs now becomes a major problem concerning social security systems. In this sense, active and appropriate ageing policies are key to attaining the mentioned balance and transforming older adults from being a "burden" to society into those being able to maintain their lives on their own since states have to maintain the health and productivity of older adults to alleviate the burden of disease and care caused by ageing and improve their welfare. To do so, communities and neighbourhoods encourageing the participation of older adults in physical and social activities should be created, thus enabling ageing in place (Eriksson, 2011; Koohsari et al., 2018). Ageing in place refers to older adults ageing in their own homes and neighbourhoods instead of a care centre (Federmeier & Kutas, 2005; Lee & Tan, 2019). Despite being a way to secure access to log-term care, staying in a care centre may mean breaking away from one's social and physical environments and being deprived of social support (Asiamah, 2021). Thus, setting up older adults' homes to meet their needs, making neighbourhoods age-friendly, and ensuring housing and neighbourhood safety may be necessary in ageing world. Age-friendly design of neighbourhoods will likely maximize access to various services and social networks among older adults. In addition, it may support the minimization of problems (e.g. social isolation, neglect, and abuse) confronted by older adults.

Neglect and abuse of older adults have become a seminal problem in modern societies as a result of increasing life expectancy and decreasing birth rates, as well as the ageing of the population. Recognized as an alarming international issue, neglect and abuse of older adults concern both older adults and service providers and policymakers (Pillemer et al., 2016). Today, the rapidly ageing population may also signal a possible increase in the cases of elder abuse globally. Even if the rate of abuse and neglect of older adults remains constant, it seems clear that the number of victims will climb globally due to the increase in the geriatric population. Population projections indicate that the number of people aged 60 and older will rise to about 2 billion by 2050 (WHO, 2021). The number of victims of elder abuse is expected to reach nearly 320 million by 2050 (WHO, 2021). Based on the World Health Organization (WHO) data from 28 countries, one out of every six older adults was abused, and only one out of 24 abuse cases were reported and filed in

2021 (WHO, 2021). The first step to prevent this problem may be to identify the boundaries of the subject and the related risk factors.

Elder abuse and neglect are a complex phenomenon that occurs in many ways, including physical, psychological, sexual, and economic abuse and neglect (Dean, 2019). It causes injury, illness, loss of functioning, social isolation, and psychological problems and generally affects society in unfavourable ways (WHO, 2002). Elder abuse is a prevalent form of interpersonal violence (Ataullahjan et al., 2019), and the occurrence or repetition of an action harming the older adult in a relationship in which they expect trust may be the diagnostic feature of the problem (WHO, 2002). Inaction towards the needs of older adults (e.g. failure to act to provide support deserved by older adults) is also an aspect of neglect and abuse of older adults. Thus, two factors stand out in neglect and abuse of older adults: (1) abuse in a relationship where the victim expects trust and (2) any action from others that causes harm or deprivation (Phelan & Ayalon, 2020). The problem can indeed occur in many forms, including physical, psychological, sexual, and economic abuse, neglect, abandonment, and a loss of reputation and respect (Kaspiew et al., 2016). Noteworthy is the fact that abuse and neglect can undermine older adults' welfare and quality of life (Jackson & Hafemeister, 2016), and abuse and neglect may occur in a nursing home, community, or the home of the individual or his or her relations. The above ideas form the basis of our analysis and the methodology on which it is based.

5.3 Methods

In this chapter, we utilize a data set extracted from the income and living conditions survey (ILCS) by the Turkish Statistical Institute (TSI) in 2021. The ILCS has been administered periodically since 2006 in Türkiye to reveal income distribution among households and citizens and generate a profile of people's living conditions, social exclusion, and poverty by income. The survey, performed within EU harmonization practices, aims to produce data on income distribution, income-based relative poverty, living conditions, and social exclusion comparable to EU countries. The ILCS covers all settlements within the borders of the Republic of Türkiye. Although all households are targeted, those defined as the institutional population living in elderly homes, nursing homes, dormitories, prisons, military barracks, hospitals, and kindergartens were not included in the survey. The sample was then selected using stratified, two-stage, and cluster sampling methods (TSI, 2021).

Social and economic policies are closely linked to community welfare. Yet, few data sets can fully explore and compare this link between EU countries. The European Union Statistics on Income and Living Conditions (EU-SILC) aims to fill this gap through data on income, living conditions, and welfare. EU-SILC is an official mechanism to monitor the progress of EU member countries in their social inclusion goals and has been widening in scope since its launch in 2003 (Arora et al., 2015).

In this regard, we analysed the mentioned data set using SPSS. In general, 65–74 years are categorized as youngest-old age, 75–84 years as middle-old age, and over 85 years as oldest-old age (Ünlü, 2019). This classification was based on the following original age groups in the data:

- Up to 64 years: individuals younger than 65 years
- 65–74 years, youngest olds
- 75–84 years: middle olds
- 85 years and over: oldest olds

EU data was extracted from the EU-SILC released by the EUROSTAT. Due to the limited age classification in the released data, we considered the following groups for the parameters of housing and neighbourhood health and safety:

- Up to 64 years: individuals younger than 65 years
- 65 years and over: older people

Again, we considered the following age groups in the EU poverty data analysed in the parameter of older adults' socioeconomic status (SES):

- Up to 64 years: individuals younger than 65 years
- 65–74 years, youngest olds
- 75 years and over: middle and oldest olds

As shown above, all parameters could not be analysed over the same age group due to the differences in the age classification of the data sets used in the analysis within the chapter. In addition, the "neighbourhood safety" parameters have been clearly revealed through the analysis, and an important projection has been provided about the safety and abuse of the elderly.

5.4 Findings

To be able to understand neighbourhood safety, we first identify the factors that may cause elder abuse in an unsafe neighbourhood or living space. For older adults, lack of safety may first start in their homes and then expand around. Poverty resulting from lack of income also creates a lack of safety for them. In this regard, we address the components of neighbourhood safety for older adults as *housing health and safety, the health and safety of the residential environment*, and *older adults' SES*. Thanks to the TSI and EUROSTAT data in each parameter, we can compare older adults' welfare between Türkiye and EU countries and analyse neglect and abuse of older adults in the mentioned countries within neighbourhood safety, revealing a remarkable analysis of the mentioned social class's living conditions and welfare. Moreover, considering the data, we propose policy recommendations regarding neighbourhood safety based on older adults' welfare levels in Türkiye.

5.4.1 Indicators for Neighbourhood Safety in Türkiye on the ILCS

The parameters indicated in Fig. 5.1, which are indicators of neighbourhood safety, are used to compare the neighbourhood safety of the elderly in Türkiye and the EU. Table 5.1 also shows data relating to these parameters.

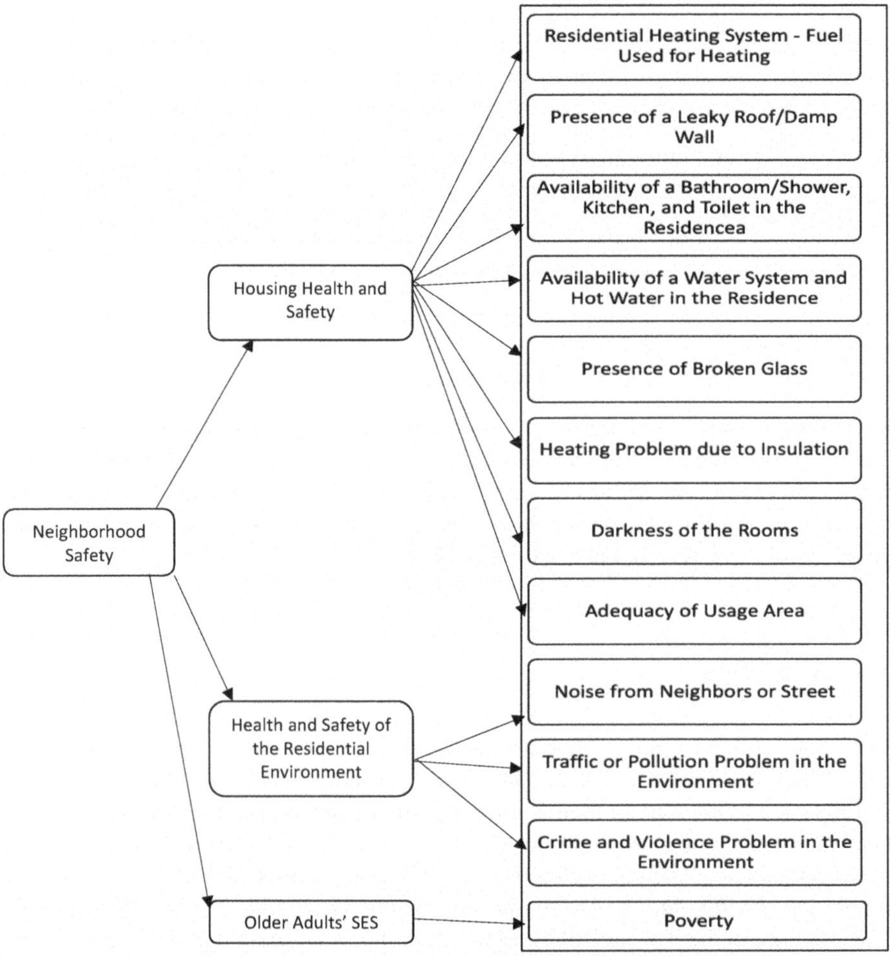

Note: The classification includes three parameters related to neghborhood safety. These are housing health and safety, health and safety of the residential environment and older adults' SES. The housing health and safety parameter consists of residential heating system – fuel used for heating, presence of a leaky roof/damp wall, availability of a bathroom/shower, kitchen, and toilet in the residence, availability of a water system and hot water in the residence, presence of broken glass, heating problem due to insulation, darkness of the rooms and adequacy of usage area. The health and safety of the residential environment parameter consists of noise from neighbors or street, traffic or pollution problem in the environment, crime and violence in the environment. Finally the older adults' SES parameter is poverty.

Fig. 5.1 Classification of the neighbourhood safety parameters

Table 5.1 Indicators for older adults' housing health and safety in Türkiye

Heating system of the residence		
Individual younger than 65 years	N	Valid percentage (%)
Heating stove	25,700,238	34.4
Radiator	46,248,018	61.9
Air conditioner	2,668,676	3.6
Other	28,369	0
Youngest olds	N	Valid percentage (%)
Heating stove	2,319,947	44.8
Radiator	2,647,309	50
Air conditioner	206,288	4
Other	2848	0.1
Middle olds	N	Valid percentage (%)
Heating stove	1,147,942	50.1
Radiator	1,062,725	46.
Air conditioner	78,248	34.4
Other	2484	0.1
Oldest olds	N	Valid percentage (%)
Heating stove	350,883	53.4
Radiator	277,942	42.4
Air conditioner	24,822	3.8
Other	2181	0.3
Availability of a bathroom or shower in the residence		
Individuals younger than 65 years	N	Valid percentage (%)
Yes, in personal use	73,982,757	99.1
Yes, in shared use	91,326	0.1
No	597,184	0.8
Youngest olds	N	Valid percentage (%)
Yes, in personal use	5,093,901	98.3
Yes, in shared use	10,488	0.2
No	79,300	1.5
Middle olds	N	Valid percentage (%)
Yes, in personal use	2,224,676	97.1
Yes, in shared use	5663	0.2
No	61,526	2.7
Oldest olds	N	Valid percentage (%)
Yes, in personal use	630,281	96
Yes, in shared use	3969	0.6
No	22,290	3.4
Availability of a toilet in the residence		
Individuals younger than 65 years	N	Valid percentage (%)
Yes, in personal use	72,138,842	96.6
Yes, in shared use	251,671	0.3
No	2,280,755	3.1
Youngest olds	N	Valid percentage (%)
Yes, in personal use	4,898,803	94.5

(continued)

Table 5.1 (continued)

Yes, in shared use	30,724	0.6
No	254,163	4.9
Middle olds	N	Valid percentage (%)
Yes, in personal use	2,134,132	93.1
Yes, in shared use	19,237	0.8
No	138,494	6
Oldest olds	N	Valid percentage (%)
Yes, in personal use	603,963	92
Yes, in shared use	6422	1
No	46,155	7
Availability of a kitchen in the residence		
Individuals younger than 65 years	N	Valid percentage (%)
Yes	74,379,163	99.6
No	292,104	0.4
Youngest olds	N	Valid percentage (%)
Yes	5,164,030	99.6
No	19,660	0.4
Middle olds	N	Valid percentage (%)
Yes	2,272,769	99.2
No	19,067	0.8
Oldest olds	N	Valid percentage (%)
Yes	647,672	98.6
No	8868	1.4
Availability of a hot water system in the residence		
Individuals younger than 65 years	N	Valid percentage (%)
Yes	71,487,568	95.7
No	3,183,700	4.3
Youngest olds	N	Valid percentage (%)
Yes	4,908,068	94.7
No	275,622	5.3
Middle olds	N	Valid percentage (%)
Yes	2,095,577	91.4
No	196,288	8.6
Oldest olds	N	Valid percentage (%)
Yes	587,281	89.5
No	69,259	10.5
Broken glass, a leaky roof, or a ruined wall in the residence		
Individuals younger than 65 years	N	Valid percentage (%)
Yes	25,305,001	33.9
No	49,366,266	66.1
Youngest olds	N	Valid percentage (%)
Yes	1,688,404	32.6
No	3,495,286	64.7
Middle olds	N	Valid percentage (%)

(continued)

Table 5.1 (continued)

Yes	808,755	35.3
No	1,483,110	64.7
Oldest olds	N	Valid percentage (%)
Yes	256,264	39
No	400,275	61
Heating problem due to insulation		
Individuals younger than 65 years	N	Valid percentage (%)
Yes	25,611,121	34.3
No	49,060,146	65.7
Youngest olds	N	Valid percentage (%)
Yes	1,727,153	33.3
No	3,456,537	66.7
Middle olds	N	Valid percentage (%)
Yes	804,519	35.1
No	1,487,346	64.9
Oldest olds	N	Valid percentage (%)
Yes	243,136	37
No	413,403	63
Problem of rooms being dark or not getting enough daylight		
Individuals younger than 65 years	N	Valid percentage (%)
Yes	12,296,639	16.5
No	62,374,629	8.5
Youngest olds	N	Valid percentage (%)
Yes	740.992	14.3
No	4,442,698	85.7
Middle olds	N	Valid percentage (%)
Yes	332,778	14.5
No	1,959,088	85.5
Oldest olds	N	Valid percentage (%)
Yes	87,235	13.3
No	569,304	86.7
Adequacy of usage area in the residence		
Individuals younger than 65 years	N	Valid percentage (%)
Yes	57,458,813	77
No	17,195,454	23
Youngest olds	N	Valid percentage (%)
Yes	4,552,926	87.8
No	630,764	12.2
Middle olds	N	Valid percentage (%)
Yes	2,036,137	88.8
No	255,728	11.2
Oldest olds	N	Valid percentage (%)
Yes	565,478	86.1
No	91,062	13.9

In Table 5.1, the group with the highest rate of heating stove is the oldest olds with 53.4%. The statistics on older adults' housing health and safety clearly show that housing safety-related problems are more common in the oldest olds in most parameters (Table 5.1). Given the distribution of the heating system by age, heating with a stove, which is also accepted as a relative poverty indicator, is the highest in the oldest olds' group (Table 5.1). In terms of the presence of essential parts in the house for humane accommodation (e.g. bathroom, shower, toilet, and kitchen), it is also evident that the absence or shared use of these parts is particularly prevalent among middle and oldest olds. More than 10% of the oldest olds do not have a hot water system (Table 5.1). Besides, the responses to the situation, "broken glass, a ruined wall, or a leaking roof in the residence", which is accepted as a fundamental indicator for housing safety, are striking. In Table 5.1, while 32.6% of the youngest olds live in unsafe houses, this rate is higher at 39% among the oldest olds. In other words, one out of three older than 65 years lives in an unsafe house. While health problems increase with age, it also applies to heating problems. It seems older adults suffer more than other age groups regarding the inadequacy of the usage area in their houses. Overall, it can confidently be proposed that older adults' housing health is worse and that their housing welfare is significantly poorer than other age groups.

The health and safety of the residential environment is another indicator for neighbourhood safety among older adults. In Table 5.2, the group with the highest noise from neighbours or street is individuals younger than 65 years. Similarly, the group with the highest rate of air pollution or other environmental problems caused by traffic or industry in the residential environment and intense crime or violence in the residential environment is individuals younger than 65 years. In this sense, our findings reveal that older adults do not become disadvantageous in the relevant parameters of the mentioned indicator, as in housing health and safety, but are exposed to noise and pollution in the environment, a significant risk factor for old age (Table 5.2). It can be implied that noise and pollution are among the factors to be alleviated to prevent elder abuse and contribute to their welfare. Although noise and pollution threaten their welfare and deepen their disadvantages, older adults do not often confront intense crime in their residential environment. One should not underestimate that the aforementioned group is more prone to diseases/disorders caused by pollution or noise than other age groups.

Poverty is a compelling significant social risk and among the greatest barriers to living a decent life. In Fig. 5.2, our findings highlight that poverty turns into an alarming danger with age in Türkiye. From 2016 to 2019, poverty among individuals aged over 85 years was higher than in all other age groups (Fig. 5.2). Similarly, poverty is exacerbated with age in all age groups. Poverty often emerges due to being deprived of income. It is acknowledged that ageing is a natural social risk in an entropic sense and that poverty becomes prevalent among older adults due to

Table 5.2 Indicators for the health and safety of older adults' residential environment in Türkiye

Noise from neighbours or street		
Individuals younger than 65 years	N	Valid percentage (%)
Yes	12,419,229	16.6
No	62,252,038	83.4
Youngest olds	N	Valid percentage (%)
Yes	657,031	12.7
No	4,526,660	87.3
Middle olds	N	Valid percentage (%)
Yes	212,704	9.3
No	2,079,161	90.7
Oldest olds	N	Valid percentage (%)
Yes	56,192	8,6
No	600,348	91.4
Air pollution or other environmental problems caused by traffic or industry in the residential environment		
Individuals younger than 65 years	N	Valid percentage (%)
Yes	17,971,211	24.1
No	56,700,057	75.9
Youngest olds	N	Valid percentage (%)
Yes	952,491	18.4
No	4,231,199	81.6
Middle olds	N	Valid percentage (%)
Yes	361,173	15.8
No	1,930,692	84.2
Oldest olds	N	Valid percentage (%)
Yes	94,324	14.4
No	562,215	85.6
Intense crime or violence in the residential environment		
Individuals younger than 65 years	N	Valid percentage (%)
Yes	7,802,759	10.4
No	66,868,508	896
Youngest olds	N	Valid percentage (%)
Yes	308,412	5.9
No	4,875,278	94.1
Middle olds	N	Valid percentage (%)
Yes	124,513	5.4
No	2,167,352	94.6
Oldest olds	N	Valid percentage (%)
Yes	26,683	4.1
No	629,856	95.9

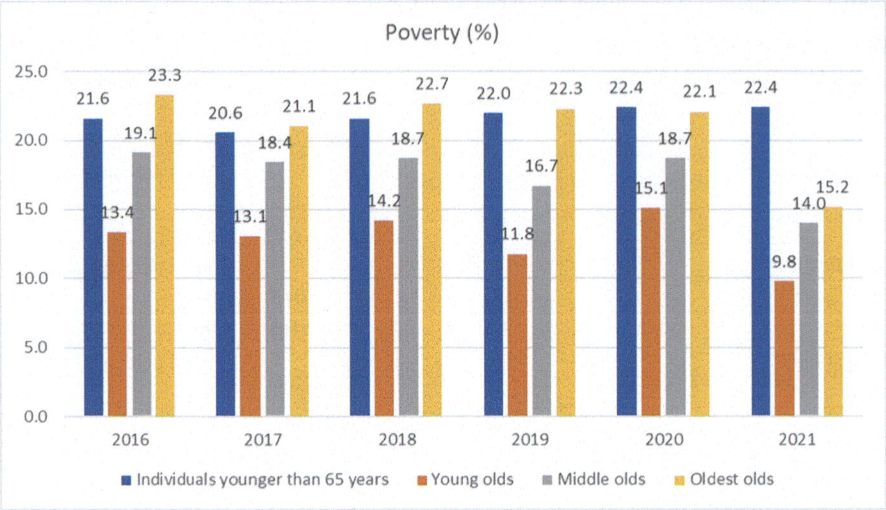

Fig. 5.2 Indicator for older adults' SES in Türkiye (%). (Adapted from: The analysis of the TSI ILSC 2016–2021 data)

their inability to work. However, this situation contradicts the social justice/security principles of social policy. Older adults deprived of adequate income should be protected from poverty more effectively with the help of direct or indirect social transfers or benefits. Our analysis of older adults' poverty, the final parameter of neighbourhood safety, suggests that poverty increases with age and that the poorest people are the oldest olds (Fig. 5.2).

5.4.2 Indicators for Neighbourhood Safety in the EU on the EUROSTAT Data

This section presents findings on indicators of neighbourhood safety using EU and the EUROSTAT data. Table 5.3 shows findings on the key parameters of neighbourhood safety as introduced earlier.

Table 5.3 indicates that the housing welfare of older European adults is considerably higher than that of those living in Türkiye. Thus, older European adults have better conditions regarding housing health and safety compared to their Turkish counterparts, highlighting a need for attaching more importance to improving housing health and safety in Turkey (Table 5.4).

When it comes to the health and safety of the residential environment, noise, violence, air pollution, and other environmental problems were higher in the EU,

Table 5.3 Indicators for older adults' housing health and safety in the EU (EU-27)

Availability of a bathroom or shower in the residence	
Individuals younger than 65 years	Valid percentage (%)
No	1.5
Older people	Valid percentage (%)
No	2
Availability of a toilet in the residence	
Individuals younger than 65 years	Valid percentage (%)
No	1.7
Older people	Valid percentage (%)
No	2.1
Broken glass, a leaky roof, or a ruined wall in the residence	
Individuals younger than 65 years	Valid percentage (%)
Yes	15.1
Older people	Valid percentage (%)
Yes	12.7
Heating problem in the residence	
Individuals younger than 65 years	Valid percentage (%)
Yes	9.1
Older people	Valid percentage (%)
Yes	9.4
Problem of rooms being dark or not getting enough daylight	
Individuals younger than 65 years	Valid percentage (%)
Yes	6.9
Older people	Valid percentage (%)
Yes	6
Adequacy of usage area in the residence	
Individuals younger than 65 years	Valid percentage (%)
Inadequate	18
Older people	Valid percentage (%)
Inadequate	6.7

compared to Türkiye (see Tables 5.2 and 4.4). Thus, older adults in the rest of EU were more vulnerable to environmental problems.

In terms of poverty, there is a similar pattern in the EU countries compared to Türkiye: poverty increases with age (Fig. 5.3). Nevertheless, poverty rates among older adults in the EU are quite different from Türkiye. While the poverty rate among older adults is about 7.5% within a 5-year period in the EU countries, it is higher at 15% in Türkiye. Hence, we can assert that older European adults' SES seems better compared to those living in Türkiye.

Table 5.4 Indicators for the health and safety of older adults' residential environment in the EU (EU-27)

Noise from neighbours or street	
Individuals younger than 65 years	Valid percentage (%)
Yes	24.7
Older people	Valid percentage (%)
Yes	14.8
Air pollution or other environmental problems caused by traffic or industry in the residential environment	
Individuals younger than 65 years	Valid percentage (%)
Yes	16.1
Older people	Valid percentage (%)
Yes	11.8
Intense crime or violence in the residential environment	
Individuals younger than 65 years	Valid percentage (%)
Yes	13.4
Older people	Valid percentage (%)
Yes	10

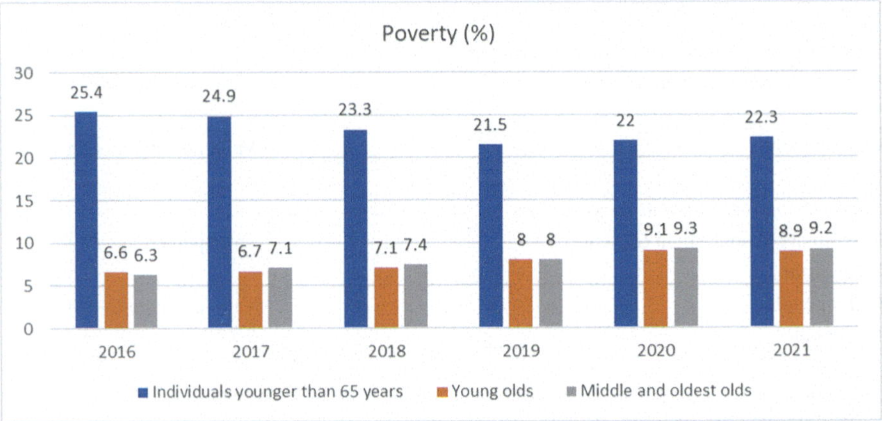

Fig. 5.3 Indicator for older adults' SES in EU (EU-27) (%). (Adapted from: EUROSTAT, https://ec.europa.eu/eurostat/data/database. Retrieval date: 02/16/2023)

5.5 Discussion

The chapter aims to explore the impacts of neighbourhood safety on neglect and abuse of older adults. This part of the chapter discusses the impacts of neighbourhood safety as a contextual risk factor on neglect and abuse of older adults.

Older adults' ageing in place depends very much on environmental factors such as safety, which is recognized as a noteworthy determinant of various health-related behaviours and outcomes (Loukaitou-Sideris & Eck, 2007). Living in disadvantaged neighbourhoods, particularly those characterized by poverty, is associated with weak social ties, problems accessing healthcare and other services, reduced physical activity, movement restrictions, and elevated stress. Disadvantaged neighbourhoods present more crime and pollution, poorer infrastructure, and restricted healthcare resources (Population Reference Bureau, 2017).

Ruelas-González and Salgado (2009) carried out a study with 799 older people in disadvantaged neighbourhoods in 2009 and found the prevalence of elder abuse to be 16.3%. Yon et al. (2017) reviewed 52 studies from 28 countries in 2017 and discovered the prevalence of global elder abuse to be 15.7%. In another meta-analysis in the same year, the prevalence of elder abuse was calculated to be 10% (Ho et al., 2017). The WHO reported the prevalence of neglect and abuse of older adults to be 15.7% in 2019 (WHO, 2019). These studies and the current chapter show that living in disadvantaged neighbourhoods may be a factor contributing to the prevalence of neglect and abuse of older adults.

A key indicator for neighbourhood safety is older adults' SES. While the literature hosts research confirming that poverty may be a risk factor for abuse (Burnes et al., 2015; Dong et al., 2007; Lachs et al., 1997; Oh et al., 2006), other studies show no relationship between income level and abuse (Kıssal & Beşer, 2009; Shugarman et al., 2003; Yan & Tang, 2004). Besides, some other studies that claim that all types of addiction contribute to abusive behaviour suggest that financial addiction is also associated with neglect and abuse of older adults (Lachs & Pillemer, 2004; Olofsson et al., 2012; Pillemer & Finkelhor, 1988; Pot et al., 1996). In addition to individual poverty, the financial status of the neighbourhood, as measured by the median household income or the proportion of people living below the poverty threshold, is accepted as an apparent indicator of older adults' welfare. Older adults living in financially disadvantaged neighbourhoods are more likely to experience chronic diseases and mobility problems and die at younger ages when compared to those living in high-income communities (Freedman et al., 2011; Grafova et al., 2008; Wight et al., 2008). Living in a disadvantaged neighbourhood is not only associated with poor cognitive functioning but also with the increased tendency of older adults to perceive their own health as bad (Glymour et al., 2010; Aneshensel et al., 2011; Wight et al., 2006). Living in high-income communities may help prevent functional problems in the early stages of disability among older people, and the opposite outcome (i.e. declining functional ability) is inherent in those living in financially disadvantaged neighbourhoods (Freedman et al., 2008).

Older people reported fewer barriers to their movements and more independent movement opportunities in areas with fewer dead ends, more connections between streets, and smooth sidewalks (Freedman et al., 2008; Grafova et al., 2008; Clarke & Gallagher, 2013). On the other hand, high crime rates and poor sidewalk designs are associated with restricted movements of older people (Gallagher et al., 2014). In their study, Boardman et al. (2012) concluded that the safety problems in the neighbourhood adversely affect older adults' cognitive functioning. Air pollution also represents another important risk factor for poor cognitive functioning among older people (Ailshire & Crimmins, 2014; Ailshire & Clarke, 2015).

5.6 Conclusion

Improving neighbourhood safety may become a milestone for ageing in place and settling older adults' problems. In this sense, it is critical to combat loneliness and social isolation, prominent risk factors for elder abuse. To do so, it is necessary to ensure older adults' active participation in society, support maintaining their social relations with friends and neighbours, and encourage their participation in social activities (e.g. visiting orphanages, volunteering, working for nongovernmental organizations or foundations). There are innovative care models such as the "senior villages" model that make ageing in place possible and emerge as an alternative to traditional institutional care. A multidisciplinary study should be carried out on how to integrate the senior village care model into countries, taking into account the number of older adults living alone, the needs of this group, financial resources, as well as cultural and social infrastructure. In line with the report obtained from this study, senior villages should be designed as a pilot application and offered to the service of older adults. Further research may need to focus comprehensively on the impacts of neighbourhood characteristics and safety on older adults, particularly on their health and elder abuse.

Additionally, relevant governing bodies may designate policies to ensure public order, a service dedicated to welfare. Police units, called the "white-uniformed" specialized in older adult care and social services, may be deployed to serve only in neighbourhoods with a predominantly older adult population. Besides, the selection of staff to serve older adults among those with fundamental knowledge of social work may also contribute to the efficiency of the welfare service to be provided. The mentioned public order service is likely to reduce the undesirable situation to arise because of the social isolation of older adults, as well as provide neighbourhood safety. Similarly, an organizational structure with working groups, called the "white desk", can be allocated to monitor and satisfy older people's safety and other needs within municipalities or governorates. For example, communication tools can be channelled to the residences of older adults in poverty or having relatively lower housing welfare to contact the "white desk" directly. Moreover, not charging older adults with court fees or assigning free lawyers in cases of potential abuse or risks endangering safety can be presented as a proposal for a deterrent policy in elder abuse.

Review Questions

1. What are the parameters that create neighbourhood security?
2. Can the phenomenon of neighbourhood safety be standardized on a global scale?
3. Is there a relationship between the phenomenon of neighbourhood safety and neglect and abuse of older adults?
4. Can actions and policies implemented to create safe neighbourhoods be beneficial in preventing neglect and abuse of older adults?
5. Can actions and policies implemented to create safe neighbourhoods promote ageing in place?

References

Ailshire, J., & Clarke, P. (2015). Fine particulate matter air pollution and cognitive function among U.S. older adults. *Journals of Gerontology, Series B: Psychological and Social Sciences, 70*(2), 322–328. https://doi.org/10.1093/geronb/gbu064

Ailshire, J., & Crimmins, E. (2014). Fine particulate matter air pollution and cognitive function among older U.S. adults. *American Journal of Epidemiology, 180*(4), 359–366. https://doi.org/10.1093/aje/kwu155

Aneshensel, C. S., Ko, M. J., Chodosh, J., & Wight, R. G. (2011). The urban neighborhood and cognitive functioning in late middle age. *Journal of Health and Social Behavior, 52*(2), 163–179. https://doi.org/10.1177/0022146510393974

Arora, V. S., Karanikolos, M., Clair, A., Reeves, A., Stuckler, D., & McKee, M. (2015). Data resource profile: The European Union statistics on income and living conditions (EU-SILC). *International Journal of Epidemiology, 44*(2), 451–461. https://doi.org/10.1093/ije/dyv069

Asiamah, N. (2021). *Socially active neighbourhoods for older adults: A consolidation of evidence from psychometric tests and cross-sectional studies*. University of Portsmouth.

Ataullahjan, A., Gaffey, M. F., Spiegel, P. B., & Bhutta, Z. A. (2019). The health impacts of displacement due to conflict of adolescents. In P. Allotey & D. D. Reidpath (Eds.), *The health of refugees-public health perspectives from crisis to settlement* (pp. 181–201). Oxford University Press.

Boardman, J. D., Barnes, L. L., Wilson, R. S., Evans, D. A., & Mendes de Leon, C. F. (2012). Social disorder, APOE-E4 genotype, and change in cognitive function among older adults living in Chicago. *Social Science & Medicine (1982), 74*(10), 1584–1590. https://doi.org/10.1016/j.socscimed.2012.02.012

Burnes, D., Pillemer, K., Caccamise, P. L., Mason, A., Henderson, C. R., Jr., Berman, J., Cook, A. M., Shukoff, D., Brownell, P., Powell, M., Salamone, A., & Lachs, M. S. (2015). Prevalence of and risk factors for elder abuse and neglect in the community: A population-based study. *Journal of the American Geriatrics Society, 63*(9), 1906–1912. https://doi.org/10.1111/jgs.13601

Clarke, P., & Gallagher, N. A. (2013). Optimizing mobility in later life: The role of the urban built environment for older adults aging in place. *Journal of Urban Health, 90*(6), 997–1009. https://doi.org/10.1007/s11524-013-9800-4

Dean, A. (2019). *Elder abuse: Key issues and emerging evidence*. Australian Institute of Family Studies.

Dong, X., Simon, M. A., & Gorbien, M. (2007). Elder abuse and neglect in an urban Chinese population. *Journal of Elder Abuse & Neglect, 19*(3–4), 79–96. https://doi.org/10.1300/J084v19n03_05

Eriksson, M. (2011). Social capital and health – Implications for health promotion. *Global Health Action, 4*(111), 4. https://doi.org/10.3402/gha.v4i0.5611

Federmeier, K. D., & Kutas, M. (2005). Ageing in context: Age-related changes in context use during language comprehension. *Psychophysiology, 42*(2), 133–141. https://doi.org/10.1111/j.1469-8986.2005.00274.x

Freedman, V. A., Grafova, I. B., Schoeni, R. F., & Rogowski, J. (2008). Neighborhoods and disability in later life. *Social Science & Medicine (1982), 66*(11), 2253–2267. https://doi.org/10.1016/j.socscimed.2008.01.013

Freedman, V. A., Grafova, I. B., & Rogowski, J. (2011). Neighborhoods and chronic disease onset in later life. *American Journal of Public Health, 101*(1), 79–86. https://doi.org/10.2105/AJPH.2009.178640

Gallagher, N. A., Clarke, P. J., & Gretebeck, K. A. (2014). Gender differences in neighborhood walking in older adults. *Journal of Aging and Health, 26*(8), 1280–1300. https://doi.org/10.1177/0898264314532686

Glymour, M. M., Mujahid, M., Wu, Q., White, K., & Tchetgen Tchetgen, E. J. (2010). Neighborhood disadvantage and self-assessed health, disability, and depressive symptoms:

Longitudinal results from the health and retirement study. *Annals of Epidemiology, 20*(11), 856–861. https://doi.org/10.1016/j.annepidem.2010.08.003
Grafova, I. B., Freedman, V. A., Kumar, R., & Rogowski, J. (2008). Neighborhoods and obesity in later life. *American Journal of Public Health, 98*(11), 2065–2071. https://doi.org/10.2105/AJPH.2007.127712
Ho, C. S., Wong, S. Y., Chiu, M. M., & Ho, R. C. (2017). Global prevalence of elder abuse: A meta-analysis and meta-regression. *East Asian archives of psychiatry: official journal of the Hong Kong College of Psychiatrists = Dong Ya jing shen ke xue zhi: Xianggang jing shen ke yi xue yuan qi kan, 27*(2), 43–55.
Jackson, S. L., & Hafemeister, T. L. (2016). Theory-based models enhancing the understanding of four types of elder maltreatment. *International Review of Victimology, 22*(3), 289–320. https://doi.org/10.1177/0269758016630887
Kaspiew, R., Carson, R., & Rhoades, H. (2016). *Elder abuse: Understanding issues, frameworks and responses*. Australian Institute of Family Studies.
Keskinoğlu, P., Giray, H., Pıçakçıefe, M., Bilgiç, N., & Uçku, R. (2004). Yaşlılarda fiziksel, finansal örselenme ve ihmal edilme. *Turkish Journal of Geriatrics., 7*(2), 57–61.
Kıssal, A., & Beşer, A. (2009). Yaşlı istismar ve ihmalinin değerlendirilmesi. *TAF Preventive Medicine Bulletin, 8*(4), 357–364.
Koohsari, M. J., Nakaya, T., & Oka, K. (2018). Activity-friendly built environments in a super-aged society, Japan: Current challenges and toward a research agenda. *International Journal of Environmental Research and Public Health, 15*(9), 2054. https://doi.org/10.3390/ijerph15092054
Lachs, M. S., & Pillemer, K. (2004). Elder abuse. *Lancet, 364*(2), 1263–1272. https://doi.org/10.1016/S0140-6736(04)17144-4
Lachs, M. S., Williams, C., O'Brien, S., Hurst, L., & Horwitz, R. (1997). Risk factors for reported elder abuse and neglect: A nine-year observational cohort study. *The Gerontologist, 37*(4), 469–474. https://doi.org/10.1093/geront/37.4.469
Lee, J. H., & Tan, T. H. (2019). Neighborhood walkability or third places? Determinants of social support and loneliness among older adults. *Journal of Planning Education and Research, 2019*, 1–14. https://doi.org/10.1177/0739456X19870295
Loukaitou-Sideris, A., & Eck, J. E. (2007). Crime prevention and active living. *American Journal of Health Promotion, 21*(4), 380–389. https://doi.org/10.4278/0890-1171-21.4s.380
Oh, J., Kim, H. S., Martins, D., & Kim, H. (2006). A study of elder abuse in Korea. *International Journal of Nursing Studies, 43*(2), 203–214. https://doi.org/10.1016/j.ijnurstu.2005.03.005
Olofsson, N., Lindqvist, K., & Danielsson, I. (2012). Fear of crime and psychological and physical abuse associated with ill health in a Swedish population aged 65-84 years. *Public Health, 126*(4), 358–364. https://doi.org/10.1016/j.puhe.2012.01.015
Öztuna, B. (2017). *Gri Tsunami'nin Sosyal Güvenlik ve İstihdam Alanındaki Etkileri, 18*. Çalışma Ekonomisi ve Endüstri İlişkileri Kongresi.
Phelan, A., & Ayalon, L. (2020). The intersection of ageism and elder abuse. In A. Phelan (Ed.), *Advances in elder abuse research* (pp. 30–49). Springer.
Pillemer, K., & Finkelhor, D. (1988). The prevalence of elder abuse: A random sample survey. *The Gerontologist, 28*(1), 51–57. https://doi.org/10.1093/geront/28.1.51
Pillemer, K., Burnes, D., Riffin, C., & Lachs, M. S. (2016). Elder abuse: Global situation, risk factors, and prevention strategies. *The Gerontologist, 56*(Suppl 2), S194–S205. https://doi.org/10.1093/geront/gnw004
Population Reference Bureau. (2017). How Neighborhoods affect the health and well-being of older Americans. *Erişim Tarihi, 02*(03), 2023. Retrieved 10 Feb 2023 from https://www.prb.org/wp-content/uploads/2020/11/TRA35-2017-neighborhoods-wellbeing-aging.pdf
Pot, A. M., van Dyck, R., Jonker, C., & Deeg, H. J. (1996). Verbal and physical aggression against demented elderly by informal caregivers in The Netherlands. *Social Psychiatry and Psychiatric Epidemiology, 31*(3–4), 156–162.

Ruelas-González, M. G., & Salgado, N. (2009). Factores asociados con el auto-reporte de maltrato en adultos mayores de México. *Revista Chilena de Salud Pública, 13*, 90–99.

Shugarman, L. R., Fries, B. E., Wolf, R. S., & Morris, J. N. (2003). Identifying older people at risk of abuse during routine screening practices. *Journal of the American Geriatrics Society, 51*(1), 24–31. https://doi.org/10.1034/j.1601-5215.2002.51005.x

Turkish Statistical Institute. (2021). *Gelir ve Yaşam Koşulları Araştırması*. Retrived 1 Feb 2023 from https://data.tuik.gov.tr

ÜnlÜ, D. (2019). *Yaşlıların Ekonomik İstismara Maruz Kalma Durumlarının İncelenmesi: İstanbul Bahçelievler Örneği, (YÜksek Lisans Tezi)*. İstanbul Sabahattin Zaim University, Social Sciences Institute, Department of Social Work.

Wight, R. G., Aneshensel, C. S., Miller-Martinez, D., Botticello, A. L., Cummings, J. R., Karlamangla, A. S., & Seeman, T. E. (2006). Urban neighborhood context, educational attainment, and cognitive function among older adults. *American Journal of Epidemiology, 163*(12), 1071–1078. https://doi.org/10.1093/aje/kwj176

Wight, R. G., Cummings, J. R., Miller-Martinez, D., Karlamangla, A. S., Seeman, T. E., & Aneshensel, C. S. (2008). A multilevel analysis of urban neighborhood socioeconomic disadvantage and health in late life. *Social Science & Medicine (1982), 66*(4), 862–872.

World Health Organization. (2002). *The Toronto declaration on the global prevention of elder abuse*. World Health Organization.

World Health Organization. (2019). *Abuse of older people*. Retrived February 5,2023 from https://www.who.int/health-topics/abuse-of-older-people#tab=tab_1

World Health Organization. (2021). *Elder abuse*. Retrived February 2,2023 from https://www.who.int/en/news-room/fact-sheets/detail/elder-abuse

Yan, E. C., & Tang, C. S. (2004). Elder abuse by caregivers: A study of prevalence and risk factors in Hong Kong Chinese families. *Journal of Family Violence, 19*(5), 269–277.

Yon, Y., Mikton, C. R., Gassoumis, Z. D., & Wilber, K. H. (2017). Elder abuse prevalence in community settings: A systematic review and meta-analysis. *The Lancet. Global Health, 5*(2), e147–e156. https://doi.org/10.1016/S2214-109X(17)30006-2

Chapter 6
Neighbourhood Services and Ageing in Place: An Extreme Industrialisation Perspective

Nestor Asiamah, Amar Kanekar, Hafiz T. A. Khan, and Pablo Villalobos Dintrans

Acronyms

CODA	Context Dynamics in Ageing
COVID-19	Coronavirus Disease 2019
NDC	Neighbourhood Development Continuum
NEWS	Neighbourhood Environment Walkability Scale
PANES	Physical Activity Neighbourhood Environment Scale
P-E	Person-Environment
SAN	Socially Active Neighbourhoods

N. Asiamah (✉)
Division of Interdisciplinary Research and Practice, School of Health and Social Care, University of Essex, Colchester, UK

Department of Gerontology and Geriatrics, Africa Centre for Epidemiology, Accra, Ghana
e-mail: n.asiamah@essex.ac.uk; nestor.asiamah@ace-gh.org

A. Kanekar
Health Education/Promotion, School of Counselling, Human Performance and Rehabilitation, University of Arkansas at Little Rock, Little Rock, AR, USA
e-mail: axkanekar@ualr.edu

H. T. A. Khan
College of Nursing, Midwifery and Healthcare, University of West London, Brentford, UK
e-mail: hafiz.khan@uwl.ac.uk

P. V. Dintrans
Programa Centro Salud Pública, Facultad de Ciencias Médicas, Universidad de Santiago, Santiago, Chile

Millennium Institute for Care Research (MICARE), Santiago, Chile
e-mail: pvillalobos.d@gmail.com

N. Asiamah et al. (eds.), *Sustainable Neighbourhoods for Ageing in Place*,
https://doi.org/10.1007/978-3-031-41594-4_6

6.1 Introduction

As population ageing intensifies, stakeholders (i.e. governments, civil society organisations, and individuals) ought to support older adults to maintain optimal health into later life. Supporting older adults to maintain health is necessary to avoid high healthcare expenditure emanating from age-related morbidity and disability. Researchers (Albert et al., 2020; Guthold et al., 2020; Smith et al., 2016) have reached a consensus that participation in social and physical activities (e.g. walking, bicycling, and recreational activities) plays a role in maintaining health across the life span. Research has also evidenced that the neighbourhood where the individual lives can encourage social and physical activities, thereby supporting health and its maintenance (Asiamah et al., 2023; Lee & Tan, 2019; Tao et al., 2021). The role of the neighbourhood in the maintenance of health has implications for "ageing in place".

Ageing in place is the maintenance of engagement, functional ability, and independence in one's home and neighbourhood rather than in a residential care home where access to contextual resources (e.g. services, parks, and green space) may be limited (Asiamah et al., 2023; Pani-Harreman et al., 2021). It includes people's sense of attachment to their home and neighbourhood, which is the psychosocial facet of ageing in place (Asiamah et al., 2023; Wahl & Gerstorf, 2018). Thus, for people to age in place, they must maintain optimal health and a sense of attachment to their neighbourhood. We admit that ageing in place has been framed as a more complicated concept (Pani-Harreman et al., 2021), but we adopt the above definition, because it constitutes the basic premise around which researchers have delineated the role of the social and physical environment in ageing.

Over the past decades, the role of the social and physical environment in health and ageing has been explained with person-environment (P-E) fit models such as the docility-proactivity hypothesis, life-space concept, context dynamics in ageing (CODA) framework, and socially active neighbourhoods (SAN) framework (Asiamah et al., 2023; Wahl & Gerstorf, 2018). These models agree that services constitute a major facet of the neighbourhood and play a role in health and ageing. Several psychometric tools developed to measure aspects of the environment (Asiamah et al., 2020, 2023; Gan et al., 2020, 2022) recognise this role and, thus, affirm the significance of services in pro-health neighbourhoods. Yet, a theoretical framework proffering how this role of services may be affected by neighbourhood-level industrial activities is unavailable. The adverse impacts of these activities on communities and their residents are being felt worldwide (Dimitriou et al., 2008; Li et al., 2018; Ntengwe, 2006) and can be expected to worsen over time. Hence, it is necessary to understand how they may attenuate the accessibility of neighbourhood services and make ageing in place less convenient.

In this chapter, therefore, we aim to develop a theoretical framework to delineate the : (1) potential adverse impact of industrial practices on the access to neighbourhood services and (2) implications for ageing in place. This chapter comprises seven main sections. The first section (i.e. the introduction) provides a background to the chapter, whereas the second section explains "age-friendliness" as a flagship concept encapsulating walkable and sociable neighbourhoods. In the third section, a theoretical framework is formulated to identify and justify the centrality of services within age-friendly contexts. We further proffer a potential change in neighbourhood services due to industrial practices in the fourth section. In the fifth section, we discuss the implications of the impact of negative industrial activities on neighbourhood services for ageing in place. Finally, we present the limitations and concluding remarks in the sixth and seventh sections, respectively.

6.2 "Age-Friendliness" as a Conceptual Pillar in Environmental Gerontology

Ageing-friendliness is generally the extent to which a neighbourhood supports the maintenance of health-seeking behaviours (e.g. physical activity and healthcare utilisation) and health over the life course (Xu et al., 2022; Zandieh & Acheampong, 2021). Since they support healthy behaviours and health, age-friendly neighbourhoods can enable people to avoid sarcopenia (i.e. a loss of bone and muscle mass) and an early decline in functional ability. There is a consensus among researchers (Asiamah et al., 2023; Marquet et al., 2017; Sallis et al., 2010) that age-friendly neighbourhoods are characterised by high residential density, street connectivity, mixed land use, and psychosocial factors (e.g. trust and safety). Street connectivity, mixed land use including commercial uses of public space in the way of service provision, and residential density are original proxies of neighbourhood walkability (Asiamah et al., 2023; Sallis et al., 2010), a flagship terminology in environmental gerontology.

Mixed land use concerns commercial and residential uses of land (Asiamah et al., 2023; Brown et al., 2013), which offer residents the opportunity to secure access to essential services. A high residential density refers to more than 25 persons or households per hectare of the gross area within the neighbourhood, excluding water bodies (Forsyth et al., 2007; Sallis et al., 2010). A high residential density accompanies a high concentration of contextual attributes (e.g. parks, streets, and social networks) and proximity to services. Research to date (Asiamah et al., 2023; Otsuka et al., 2021; Owen et al., 2007) suggests that proximity to services, which is characteristic of high walkability, encourages social and physical activities. It is, thus, understandable why many define walkability simply as how well neighbourhoods encourage walking and other active forms of transportation (e.g. bicycling, and skating). Recent advances in environmental gerontology (Asiamah et al., 2023; Gan et al., 2020), though, suggest that age-friendly contexts encapsulate more than the foregoing three domains of walkability.

Researchers contend that whether a neighbourhood would encourage walking, social inclusion and health depends, in part, on psychosocial factors such as safety, peace, and trust (Asiamah et al., 2023; Gan et al., 2022). These researchers insinuate that a context without these psychosocial factors cannot support the maintenance of health-seeking behaviours and health among older adults, especially those with physical and cognitive limitations. Given that these factors, especially peace, safety, and social cohesion, are being obliterated or weakened by negative industrial activities (Asiamah, 2022; Asiamah et al., 2020), their significance in the age-friendliness of contexts cannot be underestimated. It is worth acknowledging that measures of walkability such as the neighbourhood environment walkability scale (NEWS) and the physical activity neighbourhood environment scale (PANES) incorporate some psychosocial factors, but they exclude basic ones (e.g. safety) that would predict physical and social activities, especially among older adults (Asiamah et al., 2023). In a nutshell, the ideal place to age (i.e. an age-friendly neighbourhood) comprises walkability and psychosocial factors such as safety, social cohesion, and trust.

6.3 The Theoretical Framework

The life-space concept proposed in the 1970s (Cantor, 1975) is among the earliest works demonstrating the role of services in health-seeking behaviours, such as walking to a grocery shop. This model drew on data gathered from residents in New York to delineate factors that may encourage walking, social engagement, and health in a context. According to the concept, the availability of services and other contextual factors in a neighbourhood would encourage walking and social engagement, both of which play a central role in the maintenance of health and functional capacity over the life course (Asiamah et al., 2023; Chen et al., 2022; Jago et al., 2009). A requirement for ageing in place is the maintenance of health and functional capacity, which enables people to avoid spending the rest of their lives in nursing homes or similar residential facilities. Deductively, the concept signifies that neighbourhood services can play a significant role in ageing in place.

Decades after the life-space concept was published, researchers recognised the place of services in age-friendliness of contexts by infusing services or factors related to them (e.g. proximity to shops and institutions) in psychometric tools measuring the social and built environment. Noteworthy among these tools are the NEWS, PANEs, and ABEFs (i.e. availability of built environment factors) (Asiamah et al., 2020, 2023; Sallis et al., 2010), with the latter emphasising the role of healthcare and security services in age-friendliness. Thus, the role of services in age-friendliness has been sufficiently upheld in the measurement of the social and physical environment. Empirical evidence also suggests that the service domains of these tools are associated with social and physical activities as well as other indicators of health (Asiamah et al., 2020; Gan et al., 2020).

The CODA model (Wahl & Gerstorf, 2018) further bolsters the place of services in the social and physical environment. Like the ABEFs, the CODA recognises ageing or the maintenance of well-being across the life span as an outcome of how much residents access healthcare and other services. The SAN framework builds upon the CODA by arguing that ageing in place is an outcome of the interplay between the built environment, psychosocial factors, and personal factors (Asiamah et al., 2023). It emphasises that residents in a neighbourhood must utilise contextual resources such as services to maintain optimal health in the ageing process. Yet, individuals are unlikely to use or access services if they lack functional ability, income, and a sense of attachment to their community. A sense of attachment is implicit in psychosocial factors, such as safety and peace; people would not utilise services in a neighbourhood void of peace and safety. As observed by some researchers (Asiamah, 2022; Li et al., 2018), hazards from unfavourable industrial activities (e.g. noise, explosions, spillages, and air pollution) would obliterate psychosocial factors and subsequently hinder access to services, which we recognise in this chapter as the pillar of age-friendly contexts.

Services are not only inherent in age-friendly neighbourhoods but can also be the pivot of these contexts, since the utilisation of many other neighbourhood resources in ways that support health and ageing hinges on them. Supporting this reasoning are the CODA and SAN, which recognise that services and other contextual factors (e.g. safety, parks, and pavements) jointly influence optimal health by facilitating walking and social activities. The ecological systems theory proposed by Urie Bronfenbrenner also locates social services within an exosystem that overlaps with the macrosystem (Pérez-Wilson et al., 2021; Santos et al., 2022). While the exosystem encompasses a socioeconomic environment including neighbours and industries, the macro system concerns attitudes and community culture. The link between these two systems suggests that culture and behaviours tied to them can be due to services.

Recent analyses of P-E fit models (Asiamah et al., 2023; Wahl & Gerstorf, 2018, 2020) suggest that the recognition of services as a part of the social and built environment has stood the test of time. This recognition dates to 1951, when the first P-E fit model was developed and has progressed through the chronology of paradigms. Services have also been consistently recognised in psychometric studies since 2003 when the first walkability measure (i.e. NEWS) was published (Saelens et al., 2003), though they became more prominent in measures of the social and built environment in the last decade (Asiamah et al., 2023; Gan et al., 2022; Gan et al., 2020; Wahl & Gerstorf, 2020). This being so, it can be inferred that the recognition of services by researchers as a component of age-friendly contexts is gaining momentum.

The debate from 1951 to date suggests that an age-friendly context comprises the built environment and the social environment proxied with psychosocial factors such as trust and safety (Asiamah et al., 2023; Gan et al., 2020; Wahl & Gerstorf, 2020). In practice, the utilisation of these two broad components of the neighbourhood is influenced or facilitated by services. To illustrate, built environment attributes, such as street connectivity, traffic flow, and a high residential density, hold more meaning in service utilisation or provision. Streets, traffic signs, and a high residential density are required for using services or for accessing service providers through travel. Services may also facilitate the use of parks and other contextual factors; people who walk because the built environment offers aesthetic features may maintain their walking behaviour because of recreational services within their neighbourhood. The import of this reasoning is that people are less likely to use the built environment in the long term without services. More so, residents' maintenance of social and physical activities over the life course may not be realistic without services.

Service (e.g. healthcare) utilisation is more probable and sustainable in neighbourhoods with relevant psychosocial factors such as trust, safety, and peace since residents, especially those with physiological limitations, would not use services in unsafe neighbourhoods. In situations where services motivate residents to utilise other neighbourhood resources, psychosocial factors would play a role in sustaining behaviours such as walking and social participation over the life course (Asiamah et al., 2023). Services, therefore, serve as the pivot of age-friendly contexts that facilitate behaviours through other attributes, such as parks, social networks, and mixed land use. This viewpoint identifies a need to safeguard access to services against neighbourhood-level extreme industrial practices.

6.4 Extreme Industrialisation and Its Impact on Neighbourhood Services

We define industrialisation as a period of social and economic change during which an agrarian society evolves into an industrial society, where technology and machinery are predominantly used for the mass production of goods and services (Huang et al., 2020). Industrialisation in a neighbourhood would involve mass production of services with machinery and technologies at a scale that may be of concern to residents, but this takes time to evolve and reach extremity. In this section, possible changes in the number and variety of services over the neighbourhood development continuum (NDC) are explained.

6.4.1 Neighbourhood Service Mix and NDC

Drawing on the marketing mix concept (Abedian et al., 2022), we reason that services can only be provided in a neighbourhood at locations that people can reach. We define a neighbourhood as a 10–15 minutes walk away from a home (Saelens et al., 2003; Sallis et al., 2010) and as a group of people with a shared identity, system of beliefs and norms, as well as culture (Jenks & Dempsey, 2007). These definitions indicate that a neighbourhood includes residents as a social component and occupies a defined physical space. As a human ecosystem, a neighbourhood would start as one or a few homes with occupants who would survive by basic services, such as food vending and water supply. At its inception, though, the neighbourhood would lack most of the services residents need; hence, its few inhabitants may depend on services from nearby neighbourhoods. As the population grows over time, the "neighbourhood service mix" needed by its residents would grow (Asiamah, 2022).

Neighbourhood service mix is the variety and accessibility of services in a neighbourhood and the extent to which these services meet the needs of residents (Asiamah, 2022). A neighbourhood service mix encompasses service availability, accessibility, and relevance, all of which are recognised by researchers (Asiamah et al., 2020, 2023; Gan et al., 2020, 2022) as important factors influencing residents' use of the neighbourhood through social (e.g. recreation), economic (e.g. vending or merchandise), and physical activities. Service relevance is the extent to which a service meets the needs of residents and complements service availability to engender access to services. Table 6.1 shows the three categories of services (i.e. essential, complementary, and tertiary services) inherent in a neighbourhood service mix and their operational definitions.

Table 6.1 The authors' suggested categories of services within a neighbourhood service mix

Description	Serial number	Examples
Essential (primary) services		
These are most fundamental to survival and are daily needed or used to maintain a normal life	1	Catering and food vending
	2	Uncooked food (e.g. ingredients) vending
	3	Water sale and distribution
	4	Healthcare (clinical)
	5	Home and community care
	6	Pharmaceutical services
	7	Paramedical services
	8	Clothing and fashion services
	9	Sheltering and home maintenance
	10	Utilities for use in households
Complementary (secondary) services		
These complement essential services and are necessary for happiness, individual productivity, and a sense of security	1	Sale and distribution of home appliances
	2	Cleaning and sanitation
	3	Cooling and heating
	4	Security services (e.g. police service, fire service, etc.)
	5	Legal services
	6	Education and training
	7	Religious services
	8	Recreational services (e.g. cinemas, gyms, pubs, etc.)
	9	Transportation
	10	Beauty and grooming services (e.g. salons)
	11	Local government/council services

(continued)

Table 6.1 (continued)

Description	Serial number	Examples
Tertiary (factory-dependent) services		
These may not be directly consumed by households but are necessary for the production or supply of essential and complementary services	1	Oil refinery and supply
	2	Stone quarrying and supply
	3	Sand winning
	4	Mining and mineral processing
	5	Automobile assembling
	6	Cement production and supply
	7	Beverage manufacturing and supply
	8	Distillery and brewing
	9	Chemical (i.e. fertiliser, dies, hydrogen peroxide, etc.) engineering and supply
	10	Manufacturing and supply of plastics, paper, and cartons

Note: The three categories of services are classified based on the assumption that some services are more relevant to health and survival, especially among other adults, and are needed daily. Essential services are needed daily or weekly and are necessary for the continuity of life even in the short term. The individual can survive for a time without complementary and tertiary services. This list is not exhaustive of all services in a neighbourhood, and services are not listed in an order of relevance

Researchers (Asiamah et al., 2020, 2023; Gan et al., 2022) have suggested that essential services such as healthcare and catering are the most relevant, especially for older adults. We admit, though, that some essential services are produced or delivered with complementary and tertiary services. For instance, healthcare, water supply, and catering services depend on tertiary services such as the supply of gas and electric energy in any neighbourhood, which means that all the service categories in Table 6.1 are necessary for a healthy and normal life.

Service accessibility can be expected to increase as the human population grows, but this depends on how the neighbourhood started. Neighbourhoods usually start with one or a few homes with occupants we call early-bird settlers (Paül & Tonts, 2005; Wang et al., 2020). Such a neighbourhood's population grows to encourage the establishment of new businesses. We call this type of neighbourhood budding a "residential startup". Some other neighbourhoods emerge around a factory or group of factories set up in an isolated place (Li et al., 2018; Wang et al., 2020). Residential blocks or homes are built near the factories, often by or for stakeholders such as employees, partners, or groups with an interest in the new business. We call this type of budding an industry-driven startup, and the rest of the discussion in this chapter predominantly applies to a residential startup, which is the traditional way neighbourhoods emerge and evolve. With both startups, though, the presence of early settlers encourages other residents to settle in the neighbourhood, which means the trend of growth of the population depends on the early settlers. The chronological process by which the new neighbourhood's population grows alongside service accessibility is the NDC (see Fig. 6.1).

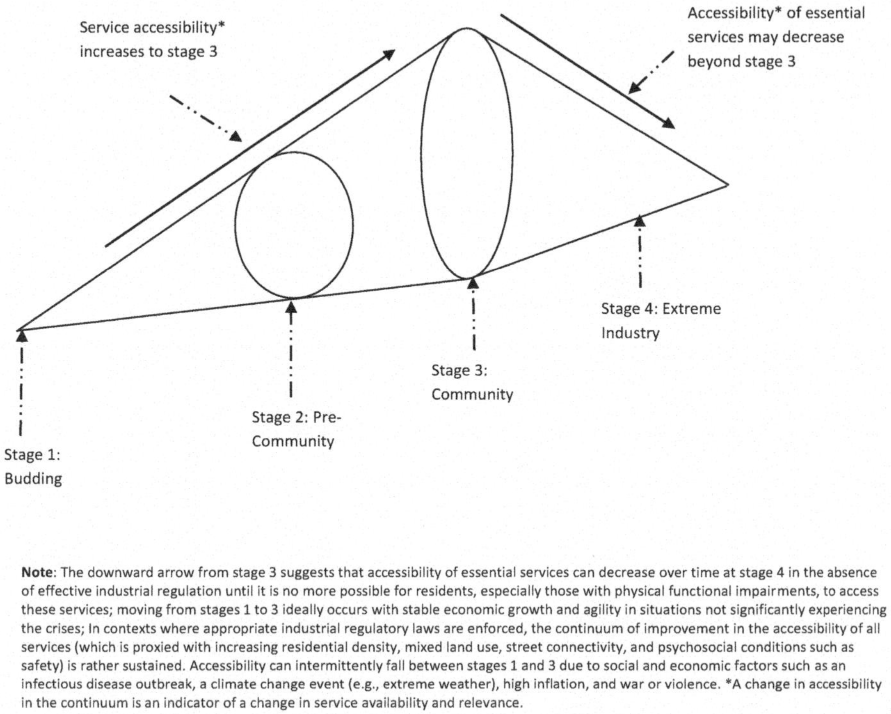

Fig. 6.1 The authors' construct of a chronological change in service mix through the neighbourhood development continuum

NDC is a continuum along which a service mix evolves with the human population of a neighbourhood. The NDC, as depicted in Fig. 6.1, starts at the budding stage of a neighbourhood, and may evolve into an "extreme industry", the stage where the absence of effective regulations encourages extreme industrial practices (Asiamah, 2022; Li et al., 2018; Ntengwe, 2006). At this final stage, residents would experience spillages, noise, explosions, and misuse of neighbourhood resources (e.g. water) due to poorly regulated industrial practices. These practices are characteristic of extreme industrialisation, which we define as the development of businesses, enterprises, or factories in a country or neighbourhood on a large scale in a way that threatens neighbourhood resources (e.g. green space, animals, and rivers), livelihoods, and the normalcy of lives in communities (Asiamah, 2022). Noteworthy at this stage is a possible decline in access to neighbourhood services, which means that the neighbourhood service mix can deteriorate from the fourth stage of the NDC due to extreme industrialisation.

In Table 6.2, we describe the four stages of the NDC and suggest their physical and social environmental attributes. Figure 6.1 and Table 6.2 suggest that the three domains of the neighbourhood service mix (i.e. availability, accessibility, and relevance) increase between stages 1 and 3 but decline from stage 4. The three

Table 6.2 Authors' operationalisation of attributes at the key stages of the neighbourhood development continuum

Stage	General attribute	Population	Typical built attributes	Typical services	Service accessibility	Availability	Accessibility	Relevance
Stage 1: Budding	Only one or a few homes with residents are found within the new neighbourhood. This is often a new site where new residential blocks or homes are being constructed	A relatively small number of residents from one or a few homes live in the neighbourhood. Population density is low	Untarred roads, non-availability of streets, low residential density, and non-availability of service delivery channels, such as malls, standard pharmacies, and banks	Food vending and the delivery of basic essential services (e.g. sale and distribution of water), with no or a few well-regulated industrial services	Only the basic essential services can be accessed, and most services can only be accessed in other neighbourhoods	Low: only a small fraction of essential services is available	Low: services may be closely located and priced, but most services needed may not be available	Low: the essential services are relevant, but most of the other relevant services are unavailable
Stage 2: Pre-community	The built environment (e.g. residential blocks, homes, streets, parks, etc.) has been completed or is nearly completed	The area is populated with residential facilities and permanent residents, a few of whom may work in available companies	Tarred roads and streets, street connectivity, high residential density, and availability of service delivery channels, such as malls and banks	All essential and complementary services, with or without a few well-regulated large-scale industrial services	Optimum access to all services through moderate pricing and service delivery channel location	Moderate: more essential services are introduced, and complementary and tertiary services may begin to emerge	Moderate: services not previously available for use and now be accessed and used	Moderate: Relevance improves since additional services supporting quality of life are introduced

Stage 3: Community	The built environment including some well-regulated large-scale factories has been completed, but residential facilities dominate the built infrastructure	The area is populated with residents, some of which work within the available factories and enterprises	Tarred roads and streets, availability of streets, high residential density, and availability of service delivery channels, including some factories and businesses	All essential services and complementary services, with many being well-regulated large-scale industrial services	High accessibility to all services through the operations of small, medium, and large-scale businesses, including well-regulated factories	High: all or most of the services (i.e. essential, complementary, and tertiary) are available	High: all or most of the services can be accessed by residents, including older adults	High: all or most of the services support quality of life
Stage 4: Extreme industry[a]	Most of the public space is occupied by unregulated or poorly regulated large-scale factories and business	The population of permanent residents is smaller, compared to the industrial stage	The completed residential infrastructure is gradually replaced with large-scale poorly regulated industrial facilities	Industrial services begin to replace essential and complementary services and eventually dominate the mix	Access to essential and complementary services begins to decline	Declines: essential and complementary services are being replaced by tertiary services	Declines: the cost of services and distance to them begins to increase	Declines: residents begin to lose access to the most important services (i.e. essential and complementary services)

[a]Extreme industrial activities may start early in the continuum (e.g. from stage 1), but these may not be intense and would, therefore, have less impact on residents and services. The intensity of extreme industrial activities may increase over time but would reach their extremity at stage 4

dimensions are low at the budding stage since most services needed by residents are unavailable, but they improve over time over the heuristic to the third stage as more services are introduced in response to increasing population and demand. This idea is supported by a principle of economics that suggests that the portfolio of services in a neighbourhood depends on demand informed by a growth in the human population (Arauzo-Carod, 2021; Butler et al., 2022). Even so, this growth occurs steadily only with appropriate macroeconomic measures and the absence of economic stressors, such as unemployment, high inflation, disease burden, and social instability from violence.

A neighbourhood at the budding stage has a few essential services typically provided by residents and petty traders (see Table 6.2). Businesses and governments focus attention on neighbourhoods with large human populations, so neighbourhoods at this stage lack services provided by these stakeholders (Wang et al., 2011). Though most services are unavailable, essential services such as water supply and food vending available are relevant as they support the survival and health of residents. Food vendors often target the few early residents and possibly individuals working on new residential construction sites who may not live in the area. The cost of services depends on the economic situation in surrounding neighbourhoods, but this is relatively low since businesses that influence the cost of goods and services are absent (Modai-Snir & van Ham, 2018; Wang et al., 2011). The distance between homes and places where services are delivered would be long, since the new neighbourhood is dispersed or less compact, and services are provided at locations that may not be accessible to every household.

The second feature of the NDC is the "pre-community" stage, the phase where most of the neighbourhood is occupied by homes or residential blocks containing residents, but the neighbourhood generally lacks attributes, such as high street connectivity, residential density, quality roads, pavements, traffic signs, parks, and other facilities found in city centres. This phase evolves into the stage of "community", where most of the neighbourhood characterises the foregoing advanced attributes of city centres or age-friendly contexts. This is the stage where age-friendliness characterised by the built attributes (i.e. mixed land use, high residential density, and street connectivity) and psychosocial factors (e.g., safety) are experienced by residents in the absence of unfavourable industrial activities. As opined by some researchers (Lee & Tan, 2019; Zandieh & Acheampong, 2021), such an age-friendly context mostly constitutes residential blocks or buildings that are served by randomly located enterprises and factories. The operations of businesses are well regulated and, therefore, have no or little adverse impact on residents.

The fourth phase is the stage of "extreme industry", where many of the residential blocks or homes that occupied public space at stage 3 have been replaced with factories or businesses providing tertiary services or performing potentially hazardous industrial activities. Occupants of residential blocks or homes left would experience noise, air pollution, and spillages. Many flats are replaced with skyscrapers or taller buildings that may be difficult to use by vulnerable groups including older adults (Asiamah, 2022; Asiamah et al., 2023). Service accessibility begins to dwindle at this stage as many of the essential services and their delivery channels are likely to be replaced with tertiary service providers (Asiamah, 2022). The age-friendliness of the neighbourhood may also begin to decline as extreme industrial practices increase the frequency of hazards, such as spillages, explosions, noise, and air pollution. Older adults are unlikely to utilise contextual resources (e.g.,, parks and services) to perform health-seeking behaviours due to the increasing likelihood of these hazards.

The NDC is a continuum of changes in the three dimensions (i.e. availability, accessibility, and relevance of services) of a neighbourhood service mix. These three attributes increase over time towards the "community stage" as the population of residents grows. Businesses get attracted to neighbourhoods as their populations grow, and their operational pattern changes with time as they leverage the increasing population to achieve high performance and scale up their infrastructure. Smaller businesses occupying smaller buildings may expand over time and take up more complex business activities such as manufacturing. Therefore, in contexts where industrial activities are poorly regulated, the activities of businesses may undermine the safety, security, and health of residents (Asiamah, 2022; Dimitriou et al., 2008; Ntengwe, 2006). This situation is depicted in the continuum with the stage of "extreme industry", where service accessibility begins to fall due to older residents' fear of the unexpected and their uncertainty about being able to use the changing built infrastructure. Though the neighbourhood can experience extreme industrial activities from its stage of budding up the heuristic, particularly if nearby neighbourhoods are already at the extreme industry stage, the adverse impact of industrial activities on residents gets stronger at the fourth stage. Extreme industrialisation at the fourth stage may provide economic benefits to the government, but it can have negative implications for access to essential services and ageing in place.

6.4.2 The Influence of Extreme Industrialisation on Access to Essential Services

As opined by researchers (Asiamah, 2022; Li et al., 2018; Ntengwe, 2006), the failure of stakeholders to regulate land ownership, use, and development results in extreme industrialisation. For example, private land developers may replace parks and other public spaces that should ultimately play an aesthetic role in the neighbourhood. Individuals and organisations may also replace smaller buildings providing essential services (e.g. domestic water supply and processing) with factories producing tertiary services. These and similar unwanted outcomes can be expected to limit access to neighbourhood services in four ways: (1) displacing individuals and businesses that provide essential services, (2) using up resources necessary for the sustainable provision of essential services, (3) making the neighbourhood less habitable, and (4) making it impossible or less possible for residents, especially those with a disability, to travel to essential service providers.

Especially in Africa where individuals and businesses can easily own landed properties (Asiamah, 2022; Ntengwe, 2006), providers of essential services (e.g.

food vendors, pharmacies and drug stops, and suppliers of drinking water) are often compelled to close down their businesses or leave their current locations either because the space within which they operate has been occupied by businesses providing tertiary services (e.g. bank) or the cost of securing this space becomes unbearable. Essential service providers who can secure their space may have no choice but to increase the cost of their services. Since service accessibility is partly determined by the cost of individual services and the proximity of places where they are delivered (Asiamah, 2022; Gan et al., 2020; Otsuka et al., 2021), any of the above situations can reduce access to essential services, especially for older adults living on a low income.

Firms practising extreme industrialisation can cut or reduce access to essential services when they use up resources that providers of essential services need to remain in business. For example, large businesses that provide tertiary services are generally more financially agile (Abedian et al., 2022; Johnson & Rasker, 1995) and, therefore, leverage their financial capability to acquire properties and spaces in the choicest of locations. Similarly, these businesses in a neighbourhood may use up most of the utilities available, such as water, electricity, and gas. Their presence in the neighbourhood can put pressure on the supply chain of these utilities and subsequently increase their cost against the low financial strength of smaller businesses that provide essential services (e.g. drug stores and food vendors). Consequently, suppliers of essential services may be unable to meet the rising cost of utilities or may run out of operational resources (e.g. electric energy, water, gas).

Researchers (Asiamah, 2022; Dimitriou et al., 2008; Li et al., 2018) have described how businesses providing tertiary services have made neighbourhoods less habitable. Noise, spillages, explosions, atmospheric dust, and movement of large vehicles (e.g. trailers) are the consequences of the operation of manufacturing companies or factories in communities. Gas explosions and spillages have necessitated the evacuation of residents from their homes and neighbourhoods in many countries (Asiamah, 2022; Li et al., 2018). Air and water pollution is a public health risk that can increase the incidence of lower respiratory diseases including asthma (Asiamah, 2022; Ntengwe, 2006). Yet, many neighbourhoods are experiencing air and water pollution due to the operational activities of factories (Asiamah, 2022; Li et al., 2018; Ntengwe, 2006). Research has shown that residents can be forced to relocate when the above industrial hazards become unbearable in a neighbourhood.

For the foregoing reasons, extreme industrial activities can limit access to essential services by discouraging travel to channels for delivering essential services (Asiamah, 2022; Jeffrey et al., 2019). Since many residents may access services only through active transportation (e.g. walking and bicycling), industrial activities that contribute to low walkability or a sense of insecurity would make it difficult for residents to reach essential services in their neighbourhoods. The movement of large commercial vehicles in the neighbourhood can also shorten the life span of tarred roads, facilitating the development of potholes on streets and roads linking to services (Asiamah, 2022). If so, any type of travel to essential services in the community can be discouraged or hindered. These possibilities have unfavourable implications for ageing in place.

6.5 Implications for Ageing in Place

Whether people would avoid sicknesses and remain engaged with life in their neighbourhoods would depend on essential services. Though tertiary services play economic and social roles in the neighbourhood (Huang et al., 2020), they should not replace essential services in the neighbourhood since they cannot support ageing without essential services. As such, extreme industrial activities that cut or reduce access to them can be viewed as a public health risk, regardless of their economic impact on the local or national economy. It is, therefore, incumbent on governments to develop and roll out policies to protect essential services. An easier way to safeguard essential services in this regard is to regulate industrial activities as well as land ownership and development within neighbourhoods, ensuring that every neighbourhood maintains a mix and balance of its services.

Industrial hazards and a lack of access to essential services can compel residents, especially older adults, to relocate to unfamiliar neighbourhoods perceived to be safer, though the working population in surrounding neighbourhoods may increase as people migrate into them to explore employment opportunities (Huang et al., 2020; Wang et al., 2020). Older residents who relocate or move into a nursing home lose a sense of attachment to their neighbourhood or the opportunity to continue ageing well at home. Thus, older residents are unlikely to meet two requirements for ageing in place, which are maintaining a sense of attachment to their neighbourhood and avoiding admission into a residential care home or a relocation to an unfamiliar neighbourhood. Hence, policies aimed at protecting or supporting older adults in neighbourhoods experiencing extreme industrialisation should be developed and enforced by governments. These efforts ought to be accompanied by routine monitoring of the activities of businesses and keeping track of the scale and type of industrial activities undertaken in neighbourhoods.

The above negative impacts of extreme industrialisation on health and ageing can exacerbate in contexts where other crises (i.e. infectious diseases, violence, and climate change events) are being concurrently experienced. In such settings, the likelihood of older adults losing their sense of attachment, moving into a residential care home, or relocating to another neighbourhood would be higher. Suffice it to say that ageing out of one's home or neighbourhood is more probable if stakeholders fail to implement measures to avoid the interplay among industrialisation, violence, infectious diseases, and climate change events (e.g. extreme weather). As reported in some studies (Asiamah, 2022; Barchielli et al., 2022), neighbourhoods already facing two or more of these threats may be facing a higher public health risk and may record higher population-level morbidity, disability, and mortality. Such neighbourhoods need interventions to eliminate the multiple threats or shield residents from their impacts.

6.6 Limitations and Future Research

Our assumptions about the stage of extreme industry are only valid in contexts where there is no or little enforcement of safe industrial practices, which means that contexts implementing the right regulatory laws can improve the desired conditions of the third stage, such as mixed land use, street connectivity, residential density, and safety. Preserving these contextual attributes, especially psychosocial factors (i.e. trust, peace, and social cohesion), over the course of industrialisation can be beneficial to older adults.

We assume that essential services are more important than tertiary services as they directly influence health, quality of life, and survival. Nevertheless, this chapter does not provide empirical evidence backing this thinking; hence, we call for research aimed at assessing the relative relevance of the three categories of services. The relevance of services can be assessed or measured by developing community-based benchmarks or assessment instruments and evaluating the extent to which the services positively influence key outcomes, such as health, ageing in place, neighbourhood social cohesion, or residents' satisfaction with their neighbourhood. Furthermore, future researchers are encouraged to examine how the three service categories influence or predict the above outcomes, including the proportions of individuals in different segments of the population who permanently relocate or move into a residential care home over a defined period.

The NDC has not been empirically tested and is based on socioeconomic dynamics in population growth, including an increase in demand and supply of services (Mohan, 1984; Wang et al., 2011). Since this model may improve an understanding of the basis by which neighbourhood service mix changes over time and affect ageing in place, we call for studies testing it. A possible way to test the model is to employ a longitudinal research design to assess how service accessibility changes with neighbourhood-level industrialisation over a period. Future research can also estimate the ratio of essential to tertiary services in the neighbourhood at the budding stage and explore its potential change along the continuum. To complement this evaluation, future researchers may investigate whether the above ratio and service accessibility affect key indicators such as ageing in place, social cohesion, and residents' satisfaction with their neighbourhood.

6.7 Conclusions

Neighbourhood services play an important role in the maintenance of health, but essential services may more strongly influence health and ageing in place. The NDC is a sequence along which access to essential services can decline. A neighbourhood may face the risk of losing essential services at any stage in the NDC, but this risk is higher at the stage of extreme industry, where there are poor regulatory laws against extreme industrial practices. Consequently, older residents may have reduced access to essential services, a lower sense of attachment to the neighbourhood, or may be compelled to relocate or move into nursing homes due to health problems caused or facilitated by extreme industrialisation. Against this backdrop, regulatory laws ought to be enacted and implemented to avoid extreme industrial activities or to ensure they do not limit access to services, especially essential services. Health promotion policies providing special support for older adults in neighbourhoods experiencing extreme industrialisation are necessary. Ideally, an evidence-based neighbourhood mix that balances essential, complementary, and tertiary services should be maintained along the NDC.

Review Questions

1. How would you define "age-friendly" neighbourhoods? What are the various factors which promote and hinder "ageing in place"?
2. Compare and contrast the roles of essential and tertiary services in supporting ageing in place.
3. Identify the various stages of the neighbourhood development continuum. Which do you feel are important stages in terms of the effect of industrialisation?
4. Describe ways in which industrialisation can affect access to essential services.
5. Create a systems-thinking model which can support the development of health promotion policies and the promotion of accessibility of essential services for "ageing in place".
6. Based on this chapter, devise some future research models to minimise the hazards of industrialisation on age-friendly neighbourhoods?

References

Abedian, M., Amindoust, A., Maddahi, R., & Jouzdani, J. (2022). A game theory approach to selecting marketing-mix strategies. *Journal of Advances in Management Research, 19*(1), 139–158. https://doi.org/10.1108/JAMR-10-2020-0264

Albert, F. A., Crowe, M. J., Malau-Aduli, A. E. O., & Malau-Aduli, B. S. (2020). Physical activity promotion: A systematic review of the perceptions of healthcare professionals. *International Journal of Environmental Research and Public Health, 17*(12). https://doi.org/10.3390/ijerph17124358

Arauzo-Carod, J. M. (2021). Location determinants of high-tech firms: An intra-urban approach. *Industry and Innovation, 28*(10), 1225–1248. https://doi.org/10.1080/13662716.2021.1929868

Asiamah, N. (2022). Walkable urban neighborhoods: The adverse effects of industrialization and climate change in developing countries. In *Sustainable urbanism in developing countries* (pp. 247–262). CRC Press. https://doi.org/10.1201/9781003131922-16.

Asiamah, N., Kouveliotis, K., Eduafo, R., & Borkey, R. (2020). The influence of community-level built environment factors on active social network size in older adults: Social activity as a moderator. *International Quarterly of Community Health Education, 41*(1), 77–87. https://doi.org/10.1177/0272684X20915379

Asiamah, N., Bateman, A., Hjorth, P., Khan, H. T. A., & Danquah, E. (2023). *Socially active neighborhoods: Construct operationalization for aging in place, health promotion and psychometric testing* (pp. 1–10).

Barchielli, B., Cricenti, C., Gallè, F., Sabella, E. A., Liguori, F., Da Molin, G., Liguori, G., Orsi, G. B., Giannini, A. M., Ferracuti, S., & Napoli, C. (2022). Climate changes, natural resources depletion, COVID-19 pandemic, and Russian-Ukrainian war: What is the impact on habits change and mental health? *International Journal of Environmental Research and Public Health, 19*(19). https://doi.org/10.3390/ijerph191911929

Brown, S. C., Pantin, H., Lombard, J., Toro, M., Huang, S., Plater-Zyberk, E., Perrino, T., Perez-Gomez, G., Barrera-Allen, L., & Szapocznik, J. (2013). Walk score®. *American Journal of Preventive Medicine, 45*(2), 202–206. https://doi.org/10.1016/j.amepre.2013.03.021

Butler, G., Szili, G., & Huang, H. (2022). Cultural heritage tourism development in Panyu District, Guangzhou: Community perspectives on pride and preservation, and concerns for the future. *Journal of Heritage Tourism, 17*(1), 56–73. https://doi.org/10.1080/1743873X.2021.1881524

Cantor, M. H. (1975). Life space and the social support system of the inner city elderly of New York. *Gerontologist, 15*(1), 23–27. https://doi.org/10.1093/geront/15.1_Part_1.23

Chen, P. W., Chen, L. K., Huang, H. K., & Loh, C. H. (2022). Productive aging by environmental volunteerism: A systematic review. *Archives of Gerontology and Geriatrics, 98*(707), 104563. https://doi.org/10.1016/j.archger.2021.104563

Dimitriou, E., Karaouzas, I., Sarantakos, K., Zacharias, I., Bogdanos, K., & Diapoulis, A. (2008). Groundwater risk assessment at a heavily industrialised catchment and the associated impacts on a peri-urban wetland. *Journal of Environmental Management, 88*(3), 526–538. https://doi.org/10.1016/j.jenvman.2007.03.019

Forsyth, A., Oakes, J. M., Schmitz, K. H., & Hearst, M. (2007). Does residential density increase walking and other physical activity? *Urban Studies, 44*(4), 679–697. https://doi.org/10.1080/00420980601184729

Gan, D. R. Y., Fung, J. C., & Cho, I. S. (2020). Neighborhood experiences of people over age 50: Factor structure and validity of a scale. *Gerontologist, 60*(8), E559–E571. https://doi.org/10.1093/geront/gnz111

Gan, D. R. Y., Mahmood, A., Routhier, F., & Mortenson, W. B. (2022). Walk/wheelability: An inclusive instrument pair for participatory age-friendly research and practice. *Gerontologist, 62*(1), E39–E47. https://doi.org/10.1093/geront/gnab079

Guthold, R., Stevens, G. A., Riley, L. M., & Bull, F. C. (2020). Global trends in insufficient physical activity among adolescents: A pooled analysis of 298 population-based surveys with 1·6 million participants. *The Lancet Child and Adolescent Health, 4*(1), 23–35. https://doi.org/10.1016/S2352-4642(19)30323-2

Huang, Y., Chen, C., Su, D., & Wu, S. (2020). Comparison of leading-industrialisation and crossing-industrialisation economic growth patterns in the context of sustainable development: Lessons from China and India. *Sustainable Development, 28*(5), 1077–1085. https://doi.org/10.1002/sd.2058

Jago, R., Brockman, R., Fox, K. R., Cartwright, K., Page, A. S., & Thompson, J. L. (2009). Friendship groups and physical activity: Qualitative findings on how physical activity is initiated and maintained among 10-11 year old children. *International Journal of Behavioral Nutrition and Physical Activity, 6*, 1–9. https://doi.org/10.1186/1479-5868-6-4

Jeffrey, D., Boulangé, C., Giles-Corti, B., Washington, S., & Gunn, L. (2019). Using walkability measures to identify train stations with the potential to become transit oriented developments located in walkable neighbourhoods. *Journal of Transport Geography, 76*(April), 221–231. https://doi.org/10.1016/j.jtrangeo.2019.03.009

Jenks, M., & Dempsey, N. (2007). Defining the Neighbourhood: Challenges for empirical research linked references are available on JSTOR for this article: Defining the neighbourhood challenges for empirical research. *The Town Planning Review, 78*(2), 153–177.

Johnson, J. D., & Rasker, R. (1995). The role of economic and quality of life values in rural business location. *Journal of Rural Studies, 11*(4), 405–416. https://doi.org/10.1016/0743-0167(95)00029-1

Lee, J. H., & Tan, T. H. (2019). Neighborhood walkability or third places? Determinants of social support and loneliness among older adults. *Journal of Planning Education and Research*. https://doi.org/10.1177/0739456X19870295

Li, M., Li, C., & Zhang, M. (2018). Exploring the spatial spillover effects of industrialization and urbanization factors on pollutants emissions in China's Huang-Huai-Hai region. *Journal of Cleaner Production, 195*, 154–162. https://doi.org/10.1016/j.jclepro.2018.05.186

Marquet, O., Hipp, J. A., & Miralles-Guasch, C. (2017). Neighborhood walkability and active ageing: A difference in differences assessment of active transportation over ten years. *Journal of Transport and Health, 7*(March), 190–201. https://doi.org/10.1016/j.jth.2017.09.006

Modai-Snir, T., & van Ham, M. (2018). Neighbourhood change and spatial polarization: The roles of increasing inequality and divergent urban development. *Cities, 82*(May), 108–118. https://doi.org/10.1016/j.cities.2018.05.009

Mohan, R. (1984). The effect of population growth, the pattern of demand and of technology on the process of urbanization. *Journal of Urban Economics, 15*(2), 125–156. https://doi.org/10.1016/0094-1190(84)90011-1

Ntengwe, F. W. (2006). Pollutant loads and water quality in streams of heavily populated and industrialised towns. *Physics and Chemistry of the Earth, 31*(15–16), 832–839. https://doi.org/10.1016/j.pce.2006.08.025

Otsuka, N., Wittowsky, D., Damerau, M., & Gerten, C. (2021). Walkability assessment for urban areas around railway stations along the Rhine-alpine corridor. *Journal of Transport Geography, 93*(July 2020), 103081. https://doi.org/10.1016/j.jtrangeo.2021.103081

Owen, N., Cerin, E., Leslie, E., duToit, L., Coffee, N., Frank, L. D., Bauman, A. E., Hugo, G., Saelens, B. E., & Sallis, J. F. (2007). Neighborhood walkability and the walking behavior of Australian adults. *American Journal of Preventive Medicine, 33*(5), 387–395. https://doi.org/10.1016/j.amepre.2007.07.025

Pani-Harreman, K. E., Bours, G. J. J. W., Zander, I., Kempen, G. I. J. M., & Van Duren, J. M. A. (2021). Definitions, key themes and aspects of "ageing in place": A scoping review. *Ageing and Society, 41*(9), 2026–2059. https://doi.org/10.1017/S0144686X20000094

Paül, V., & Tonts, M. (2005). Containing urban sprawl: Trends in land use and spatial planning in the metropolitan region of Barcelona. *Journal of Environmental Planning and Management, 48*(1), 7–35. https://doi.org/10.1080/0964056042000308139

Pérez-Wilson, P., Marcos-Marcos, J., Morgan, A., Eriksson, M., Lindström, B., & Álvarez-Dardet, C. (2021). "A synergy model of health": An integration of salutogenesis and the health assets model. *Health Promotion International, 36*(3), 884–894. https://doi.org/10.1093/heapro/daaa084

Saelens, B. E., Sallis, J. F., Black, J. B., & Chen, D. (2003). Comité Européen des Entreprises Vins (CEEV) (2016b) Internal market and food safety. Comité Vins, Brussels. http://ceev.eu/policy-dossiers/internal-market-food-safety. Accessed 02 Nov 2016. *American Journal of Public Health, 93*(9), 1552–1558. http://ceev.eu/policy-dossiers/internal-market-food-safety

Sallis, J. F., Kerr, J., Carlson, J. A., Norman, G. J., Saelens, B. E., Durant, N., & Ainsworth, B. E. (2010). Evaluating a brief self-report measure of neighborhood environments for physical activity research and surveillance: Physical activity neighborhood environment scale (PANES). *Journal of Physical Activity and Health, 7*(4), 533–540. https://doi.org/10.1123/jpah.7.4.533

Santos, M., Moreira, H., Cabral, J. A., Gabriel, R., Teixeira, A., Bastos, R., & Aires, A. (2022). Contribution of home gardens to sustainable development: Perspectives from a supported opinion essay. *International Journal of Environmental Research and Public Health, 19*(20), 1–26. https://doi.org/10.3390/ijerph192013715

Smith, L., McCourt, O., Sawyer, A., Ucci, M., Marmot, A., Wardle, J., & Fisher, A. (2016). A review of occupational physical activity and sedentary behaviour correlates. *Occupational Medicine, 66*(3), 185–192. https://doi.org/10.1093/occmed/kqv164

Tao, Y., Zhang, W., Gou, Z., Jiang, B., & Qi, Y. (2021). *Planning Walkable Neighborhoods for " Aging in Place ": Lessons from Five Aging-Friendly Districts in Singapore.*

Wahl, H. W., & Gerstorf, D. (2018). A conceptual framework for studying COntext dynamics in aging (CODA). *Developmental Review, 50*(August), 155–176. https://doi.org/10.1016/j.dr.2018.09.003
Wahl, H. W., & Gerstorf, D. (2020). Person-environment resources for aging well: Environmental docility and life space as conceptual pillars for future contextual gerontology. *Gerontologist, 60*(3), 368–375. https://doi.org/10.1093/geront/gnaa006
Wang, D. T., Zhao, S. X., Gu, F. F., & Chen, W. Y. (2011). Power or market? Location determinants of multinational headquarters in China. *Environment and Planning A, 43*(10), 2364–2383. https://doi.org/10.1068/a448
Wang, X., Shi, R., & Zhou, Y. (2020). Dynamics of urban sprawl and sustainable development in China. *Socio-Economic Planning Sciences, 70*(August 2019), 100736. https://doi.org/10.1016/j.seps.2019.100736
Xu, J., Chen, Y., Wang, Y., Gao, J., & Huang, L. (2022). Association between age-friendliness of communities and frailty among older adults: A multilevel analysis. *International Journal of Environmental Research and Public Health, 19*(12). https://doi.org/10.3390/ijerph19127528
Zandieh, R., & Acheampong, R. A. (2021). Mobility and healthy ageing in the city: Exploring opportunities and challenges of autonomous vehicles for older adults' outdoor mobility. *Cities, 112*(December 2020), 103135. https://doi.org/10.1016/j.cities.2021.103135

Part II
Interventions Against the Crises

Chapter 7
Ageing in Place and Built Environment Amenities at Neighbourhood Scale: The Case of South Australia

Alpana Sivam, Sadasivam Karuppannan, and Ali Soltani

Acronyms

ABS	Australian Bureau of Statistics
ALUM	Australian Land Use and Management
ANOVA	Analysis of Variance
AURIN	Australian Urban Research Infrastructure Network
CBD	Central Business District
HM	High Amenity, Middle Old
LGAs	Local Government Areas
LISA	Local Indicators of Spatial Association
LY	Low Amenity, Youngest Old
MO	Middle Old
MY	Medium Amenity, Youngest Old
OO	Oldest Old
POI	Point of Interest
UN	United Nations
YO	Youngest Old

7.1 Introduction

Populatiown ageing is a global phenomenon. In 2020, there were 723 million people in the world aged 65 years or over (United Nations, 2020). This number is projected to double to 1.5 billion in 2050 (United Nations, 2020). Australia's demographic landscape is changing with a rapid increase in the rate of ageing and

A. Sivam (✉) · S. Karuppannan · A. Soltani
Department of Urban and Regional Planning, UniSA Creative, University of South Australia, Adelaide, SA, Australia
e-mail: Alpana.Sivam@unisa.edu.au; Sadasivam.Karuppannan@unisa.edu.au; Ali.Soltani@unisa.edu.au

N. Asiamah et al. (eds.), *Sustainable Neighbourhoods for Ageing in Place*,
https://doi.org/10.1007/978-3-031-41594-4_7

the older adult population. In 2017, 15% of Australians (3.8 million) were aged 65 and over, and it is projected to grow steadily over the coming decades, and by 2097, 12.8 million people (25%) will be aged 65 and over (ABS, 2017). As the population of older adults continues to grow, it poses a significant challenge for many countries to plan for appropriate housing options and supportive services for an ageing population (Mulliner et al., 2020; Abramsson & Andersson, 2016; Bonvalet & Ogg, 2008; de Jong & Brouwer, 2012). As people age, available choices for living arrangements may narrow as they become less able to cope with the everyday demands of living (Shrestha et al., 2020, Sivam & Karuppannan, 2007, 2008).

There is a growing interest in the concept of ageing in place among policymakers and researchers to address these societal changes. Ageing in place refers to supporting people to age in their own homes and communities with some level of independence (Rogers et al., 2020; Veeroja, 2018). Many older adults prefer to 'age in place' (Burton et al., 2011) to maintain their independence, networks, and connection to family and friends (Lowen et al., 2015, Wiles et al., 2012). These factors tend to enhance their overall health, well-being, and quality of life. As stated in Veeroja (2018, p. 1) '65% of people aged 50 years and above intended to stay in their places of residence as they aged'.

Older adults are generally confined to their home and immediate surroundings around their homes (Veeroja, 2018) compared to younger population. Amenities available in their neighbourhoods are likely to be important for them to age in place. If they are mobile and can interact with communities, it will reduce their loneliness. These interactions will increase older adults' overall well-being and quality of life (Veeroja, 2018) and hence enhance their choice of ageing in place.

Although many new concepts for housing facilities have been developed, older adults are continuously looking for options that will allow them to maintain independence, and personal control over their routines (Peace et al., 2011; Abramsson & Andersson, 2016), and allow them to remain in their homes and communities as they age even when they experience physical and cognitive declines (Robinson et al., 2020). Older people are looking to downsize by moving to smaller accommodation but do not want to lose connection with previous place (Eshelman & Evans, 2002). By providing suitable housing with supportive amenities for ageing population, quality of life could be improved and possibly reduce the demand for aged-care services (Abramsson & Andersson, 2016). To promote ageing in place, governments often have focused on improving care delivery models, and less attention is given to how the built environment can enhance well-being of aged people (Gan, 2017), by supplying accessible facilities in cost efficient manner. A key factor for ageing in place is not only diverse housing options but also it requires good built environment amenities for ageing well (Gan, 2017). Built environment professionals and policymakers face the challenge of designing new and redeveloped areas that support ageing in place by providing health, public, and community facilities within a reasonable distance of older adults' residences. While previous research has highlighted the importance of dwelling improvement and typology for successful ageing in place, the impact of location and accessibility of amenities on this choice has been overlooked. By recognising the significance of built environment amenities in enabling older adults to age in place, policymakers can make informed decisions that promote healthy and independent living for seniors.

The aim of this chapter is to assess the accessibility of older adults' cohorts to built environment amenities at the fine-grain neighbourhood scale in metropolitan Adelaide, South Australia, to investigate potentials of ageing in place in various local government areas (LGAs). This research aims to use a case study approach and GIS mapping to analyze the location and accessibility of various facilities in order to support ageing in place. While previous studies have focused on a limited number of destinations, such as healthcare facilities, parks, and community centres, this research takes a more comprehensive approach by examining multiple destinations in the areas of health, shopping, recreation, and transportation. Conceptual framework and identification of built environment amenities required for successful ageing in place are presented in Sects. 7.2 and 7.3, respectively. The method of this study is outlined in Sect. 7.4, and Sect. 7.5 presents the analysis and results. Finally, Sect. 7.6 discusses the findings and its contribution to the field.

7.2 Conceptual Framework and Study Focus

Environmental gerontologists claim that as individuals age, they tend to form stronger emotional connections to their place of residence while also becoming more susceptible to the influence of their social and physical surroundings (Lecovich, 2014). The environmental docility hypothesis (Lawton & Simon, 1968) supports this claim by suggesting that the environment's influence increases as an older person's functional abilities decrease. Furthermore, the competence-environmental press model, introduced by Lawton and Nahemow (1973), emphasizes the critical role of personal competencies and the social and physical environment in determining an individual's ability to age-in-place. Therefore, for successful ageing in place, it is crucial for both the immediate and neighbourhood environments to be supportive of independent functioning for older adults. However, different groups of older individuals may respond differently to environmental changes, potentially affecting their ability to age in place successfully.

The process of aging is a complex interplay of physiological, behavioural, social, and environmental changes that occur both at the individual and community levels. The ecological theory posits that the interaction between personal competence and the environmental pressure of the built environment in the neighbourhood impacts the well-being of older adults (Vine et al., 2012). As such, the ecological model is an appropriate framework to describe and understand the role of amenities in facilitating successful aging-in-place.

Numerous gerontological studies have discussed various barriers that hinder a person's ability to age in place, which means remaining in their home with a desired level of independence throughout later stages of life. These barriers predominantly include the physical environment, accessible community services, and social exclusion or inclusion (Wiles et al., 2012). Ageing in place is affected by three factors namely, individual circumstances, built environment amenities, and housing options.

Individual circumstances, such as socioeconomic condition, health, and confidence, play a significant role in choosing the aging-in-place option. The second factor is the availability of built environment amenities at neighbourhood level. Broadly, built environment factors are location and accessibility to amenities. Factors that influence happiness keep changing with peoples age (de Boer et al., 2017). As individual age, factors like home adaptability, infrastructure accessibility, community services, and mobility impact their daily lives. The built environment supports people in carrying out their daily activities, enhances social inclusion, and provides accessibility to necessary facilities (Bowling, 2005). Third is housing options that allows older adults to live in their own neighbourhood.

The scope of this chapter is limited to availability and accessibility of amenities to understand suitability and unsuitability of neighbourhoods for ageing in place. Based on earlier research, this chapter has identified six dimensions, namely, shopping facilities, primary health services, public space and recreation, community facilities, mobility, and natural environments. Rationales for six dimensions identified are presented in Sect. 7.3.

7.3 Identification of Built Environment Amenities

The literature review has identified a range of built environment amenities that have a significant impact on an individual's ability to age in place (refer Table 7.1). Ideally, these amenities should be located within a walkable distance of 400–800 m from an older person's home (El-Geneidy et al., 2014; Marsh et al., 2006), to enhance their confidence and ability to age in place. Table 7.1 provides a list of identified amenities and a rationale for selecting the specific amenities.

7.4 Methods

Limited knowledge exists on the distribution of amenities and accessibility for older adults in the Adelaide metropolitan area. Most of the studies are qualitative and based on limited case studies. This study aims to fill this gap by using large-scale statistical data on the distribution of amenities in the region. Unlike conventional research that relies on narrow categories of land use activities, this study uses a broader analytical framework based on large data sets of amenities. This section presents a case study, methodology, and data used to establish the framework for analyzing accessibility.

Table 7.1 Identified built environment amenities for ageing in place

Amenities	Key factors	Remarks
Shopping	Shopping centre, grocery stores, specialty shops, fruits and veggie shops, butchers, and vending machines	Access to shopping facilities improves the physical and mental well-being of older adults (Wong et al., 2021; Gariepy et al., 2015; Day, 2008)
Services	Hospital, health centre, chemist, doctors, dentist, allied health facilities, veterinary, public toilet, post office, post box, bank, ATM	The availability, location, and accessibility of health services and facilities are crucial for enhancing the health, mental well-being, and confidence of older adults to age in place (Ryser & Halseth, 2012). In addition to meeting basic needs, public facilities also provide opportunities for social inclusion (Cheng et al., 2020; Zhang et al., 2021)
Recreation	Gym, sports club, swimming pools, gaming centres, cafes, bars, restaurants, public open spaces	Public spaces and recreation facilities are crucial for promoting physical and mental well-being (Zhang et al., 2020) and fostering social inclusion (Gehl, 1987, Karuppannan & Sivam, 2011). They serve as a 'third place', where individuals can interact informally and formally, providing opportunities for social engagement. (Cuthbert & Dimitriou, 1992; Corbett & Corbett, 2000; Chan & Lee, 2007; Karuppannan & Sivam, 2011; He et al., 2022)
Mobility	Public transport hub and stops, taxi station, parking space	Creating a built environment that is age-friendly, with safe, affordable, and convenient transportation options, is essential for improving the quality of life of older adults (Wong et al., 2021; He et al., 2022; Soltani et al., 2018; Day, 2008)
Natural Environments	Parks, community gardens, nature reserves, beaches, open green space, picnic sites	Access to green and blue areas has been shown to relieve stress and mental fatigue (Nath et al., 2018) and is recognised as an important restorative resource for urban residents (Van den Berg et al., 2016). Research has also demonstrated that older adults' physical and mental well-being is linked to their ability to access these natural areas (Wong et al., 2021; Dempsey et al., 2018, Finlay et al., 2015)
Community facilities	Libraries, churches, temples, community centres, social club, community centre, cemetery	Research has shown that older adults' physical health and mental well-being are linked to their access to community facilities, which can serve as a 'third place' for social inclusion (Wong et al., 2021; Beard & Petitot, 2010; Day, 2008)

7.4.1 Case Study

This study focuses on the metropolitan Adelaide area due to limited knowledge on the distribution of amenities and accessibility for older adults in the region. Adelaide, the capital of South Australia, has a population of over 1.4 million people and the highest proportion of older adults on mainland Australia, with more than 37% of the total population aged 50 or over (BISA, 2018). The region consists of 18 local government areas (LGAs), as depicted in Fig. 7.1, which shows the density of older adults at the suburb level. Brown indicates higher density, while yellow indicates lower density. Older adults are distributed randomly throughout the region, with larger concentrations in the central, inner eastern, and inner western suburbs.

7.4.2 Data Collection and Analysis Approach

The triangulation method was used to collect and analyze data from multiple sources and methods, such as data sets and spatial analysis, to gain a better understanding of the distribution of older adults and utilities in a metropolitan area. This method can ultimately help policymakers and researchers make more informed decisions about how to allocate resources and address the needs of this population.

7.4.2.1 Data Set

The data used for this research are from secondary open-access sources. However, plenty of data mining techniques, including data cleaning, manipulation, filtering, classification, and outlier detection, are applied to make the data ready and reliable for analysis and visualisation. The first source used is the ABS Census of Population and Housing data, which includes census geography (map layers) as well as socio-economic statistics. The 2016 census data is used because when we conducted this research, the 2021 census was not published. This chapter focuses on the older adults (>65) group, which has a smaller share than the children and young adolescents (<15 years), and the working-age population (15–64) groups in Australia (ABS, 2017). The meshblock is the smallest spatial unit of analysis (n > 19,000) in the census data, and it was used in this study for socio-demographic analysis.

The second large-scale data set used for this research is the point of interest (POI) data gathered from OpenStreetMap (an open-source collaborative geospatial data source) (AURIN, 2021). POI is user-collected data that includes the record of locations on a map that someone finds useful or interesting (Quadrant, 2022). POI data consists of three types of information: name, classification, and coordinates (longitude and latitude). Google Maps, Map World, and Baidu Map are among the POI providers. Compared to large-scale land use data with limited categories as defined by the Australian Land Use and Management (ALUM), the POI data is fine-grained and has very detailed attributes of urban land use. There have been some efforts in this area to classify urban land use based on big data streams from smart devices (Pei et al., 2014; Toole et al., 2012). The shortage of POI data is that, unlike professional survey of land use types, the POI data is user-generated. On the positive side, it offers an alternative to observe and interpret urban land use from the

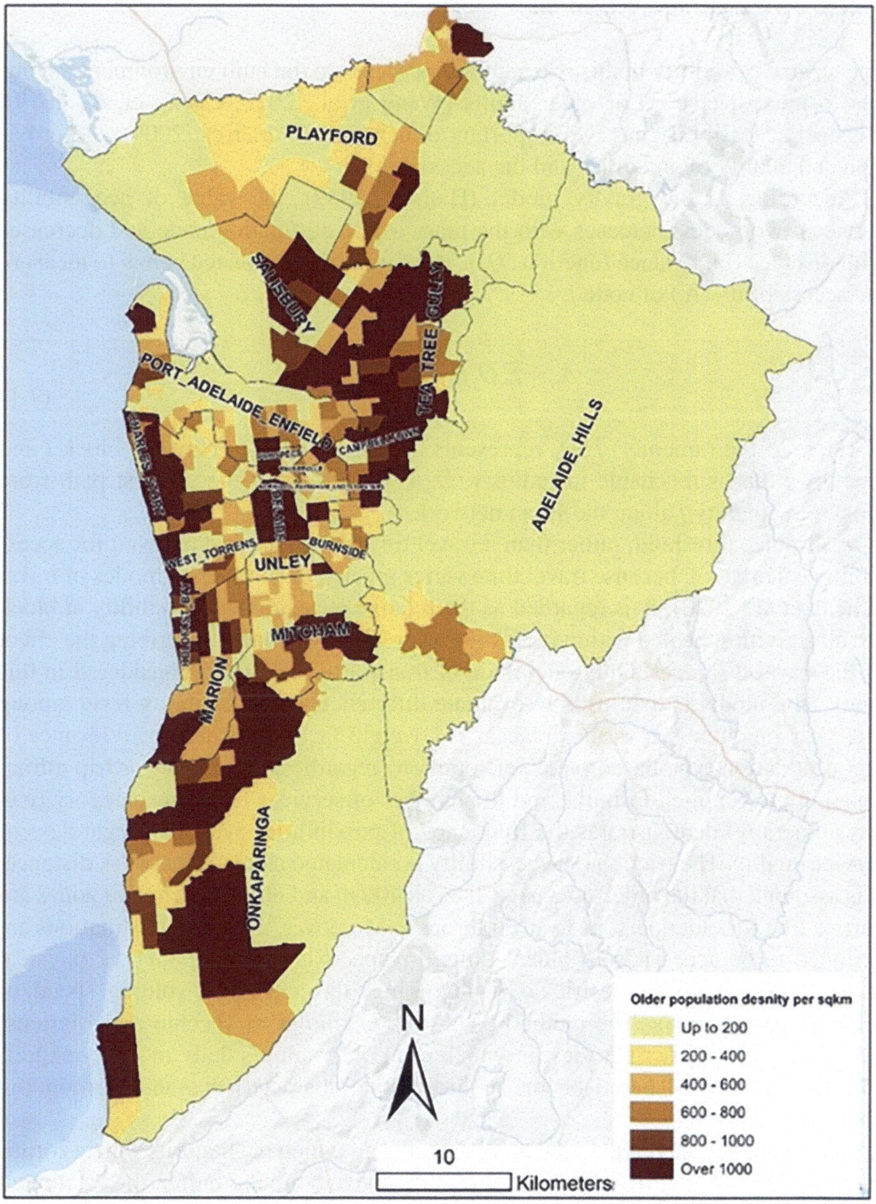

Fig. 7.1 Density of older adults in Adelaide metropolitan area, 2016. (Source: OpenStreetMap – Points of Interest (Australia) 2020; accessed from AURIN. https://data.aurin.org.au)

perspective of actual users through a bottom-up approach (Sala et al., 2020; Chen, 2015). Since POI data has approximately 50 item types, they are grouped into six main categories, as discussed in Table 7.1.

7.4.2.2 Measuring Accessibility

The term accessibility in this research refers access to the built environment amenities to meet the needs of older adults (Wang et al., 2022; Cheng et al., 2019). Physical accessibility has two important dimensions (Gutiérrez, 2009): the provision and quality of amenities and the accessibility.

According to the gravity model (Hansen, 1959), the value of accessibility between two nodes increases with the mass of the destination node and decreases with distance (impedance function '*D*'). The formula is suggested below to measure the accessibility (A_i) of node *i*:

$$A_i = \sum_{n}^{j=1} O_j D_{ij}^{-2}, \qquad D_{ij} < r \tag{7.1}$$

O_j refers to the (amenity) *j*; D_{ij} represents the travel distance between block *i* and amenity *j* that is reachable in radius *r*. D_{ij} is the length of the shortest path from block *i* to amenity *j* along the street network.

A distance threshold, rather than a travel time threshold, is employed for accessibility calculation, because travel time varies greatly, depending on modes of travel (Cheng et al., 2020). A_i is regarded as the relative measure of accessibility at block *i* to all activities located within radius *r*. Power 2 is an exponent describing the effect of the travel distances. Only one mode of transport (walking) is considered in this study. The model is then used to evaluate differences in accessibility level among older persons living in various places. The weight factor indicates the relevance of that destination type based on expert opinion (regardless of number or trip attractions) in the supplied formula, and the number of services from one category (e.g. physiotherapy) demonstrates the frequency of possibilities available, regardless of service quality. The friction of accessibility is calculated using the network distance. It is assumed that the block area as well as its 400 m and 800 m catchment zones are barrier-free for older persons to stroll through. Moreover, the stated destinations are assumed to be accessible to older adults, irrespective of age, gender, or physical ability, and access is not restricted, owing to high through-traffic volume, speed, or other safety concerns. The method used here is similar to Vecchio and Martens' (2021) notion of 'micro-stories, in which each older adult's daily mobility experience is documented based on his or her unique characteristics and surrounding context.

A catchment area refers to the areal extent from which the majority of trips originated (Dolega et al., 2016). Instead of plotting a fixed circular buffer, the catchment area is based on all walkable streets, as shown in Appendix 7.1. In this case, streets within an impedance of 400 m for 5-min and 800 m for 10-min walking (Hansen et al., 2011) are considered. The catchment polygons are more realistic and are valid where patrons are expected to use the closest facility (Dolega et al., 2016).

7.4.2.3 The Age Segmentation

The ABS census data was used to determine the areas of metropolitan Adelaide with higher proportions of older adults (65 or over). The share of older adults varies between 12.0% (Playford) and 20.0% (Holdfast Bay). We followed Choi et al.'s (2021) population segmentations with a little alteration owing to contextual differences. For each LGA, three age groups are compared. If the proportion of 65–75 is more than two other groups, it is referred to as 'youngest-old (YO)' (e.g. Adelaide); LGAs with a dominant age group of 75–85 are referred to as 'middle-old (MO)' (e.g. Marion); and if the age group of 85 is dominant, it is referred to as 'oldest-old (OO)' (e.g. Burnside). The results are provided in Appendix 7.2.

7.4.2.4 The Spatial Cross-Correlation

The spatial cross-correlation analysis approach`(Chen, 2015) was used to display the simultaneous variation in values of two variables: the density of older individuals and the amenity location. The correlation coefficient of LISA (local indicators of spatial association) and pseudo p-value were computed using GeoDa 1.20 (Anselin, 1995).

7.4.2.5 The Ability of LGAs to Support Ageing in Place

Place attachment contributes to the preference of older adults for ageing in place. Therefore, predictions of population and amenity growth over a 30-year horizon were made to assess how well LGAs satisfy ageing in place criteria. These projections assisted to understand where amenities will be required to improve in future. Appendix 7.3 provides a mathematical approach for projecting population change (Jeger, 1984; Brauer & Castillo-Chavez, 2010), and Appendix 7.4 shows formulae used. Linear, logarithmic, polynomial, exponential, and logistic models were considered, and R^2 determines which model to use in each LGA. Amenity projections are based on land use trend from 2011 to 2016. It is expected that amenities will keep growing at the same rate until 2046. Next section will describe result.

7.5 Analysis and Results

7.5.1 Population Profile and Projection for the Next 30 Years

Table 7.2 shows the 2016 population of the 18 LGAs considered in this study (ABS, 2017) and the growth for 30 years. Using the previously stated mathematical models, we projected the population for the next 30 years, as well as the proportion of

Table 7.2 Population in 2016 and projections for 30 years by local government areas

LGA	Population 2016[a]	Population growth (30 years)[b]	Population 2046
Adelaide	3037	0.864	7332
Adelaide Hills	6741	0.588	10,707
Burnside	9608	0.578	15,163
Campbelltown	10,352	0.818	18,824
Charles Sturt	21,291	0.959	41,706
Holdfast Bay	8497	0.695	14,398
Marion	15,332	0.819	27,890
Mitcham	12,534	0.617	20,266
Norwood	6909	0.745	12,055
Onkaparinga	28,278	1.166	61,255
Playford	10,790	1.552	27,532
Port Adelaide	18,964	1.029	38,480
Prospect	2807	1.512	7051
Salisbury	18,840	1.103	39,626
TT Gully	17,356	0.955	33,928
Unley	7301	1.108	15,389
Walkerville	1662	0.593	2648
West Torrens	10,296	0.465	15,082

Source: [a]ABS (2022), [b]DTI (2021)

older adults. The projections are used to compute the growth rates for the general public and the older adults between 2016 and 2046. In Greater Adelaide, Onkaparinga (169,368) has the largest population, followed by Salisbury (140,370), Port Adelaide Enfield (123,994), and Charles Sturt (114,977).

The share of older adult population range between 12% (Playford) and 23% (Holdfast Bay). The eastern and southern LGAs in many cases consist of undulating and hilly topography that would impact older adults walking, and therefore, it is expected these LGAs might accommodate lower share of older people. As shown in Table 7.2, approximately one-fifth of the metropolitan population (17%) is aged 65 or older.

According to the prediction for the next 30 years, this proportion of population will continue to increase to 22%, which is significant. The distribution of old adults will be different for various LGAs. Between 2016 and 2046, the City of Adelaide will have the highest growth rate (130.9%) in population aged over 65. Although the total population of the City of Adelaide is lower in comparison to some of the LGAs, such as Onkaparinga, Playford, etc., the growth rate for Adelaide is much higher than theses LGAs. This might be because the number of inner-city projects and massive apartment complexes has increased in the recent decade due to densification policies and excellent access to public services. Because of excellent services and variety of housing provisions, Adelaide has the potential to attract more older adults' cohort.

The LGAs surrounding the CBD can be divided into two groups: a) those with modest growth, such as West Torrens (15.4%), Burnside (32.9%), and Walkerville (38.0%), and b) those with high growth, such as Unley (58.3%), Prospect (88.0%), Norwood Payneham and St. Peters (51.2%), and Charles Sturt (66.4%). Zoning restrictions, particularly in heritage zones where large-scale construction opportunities are limited, are probably the cause for lesser growth in the inner suburbs, for example, Burnside. LGAs, such as Prospect and Norwood Payneham and St. Peters, on the other hand, could be popular locations for older people looking to relocate from the outer suburbs due to access advantages and availability of several types of housing. Existing literature demonstrates downsizing among older Australians is influenced by access to services and facilities rather than finding only smaller house (James et al., 2020).

Councils located in outer parts of the Adelaide metropolitan area, such as Playford (108.1%), Port Adelaide Enfield (71.8%), Onkaparinga (69.0%), Salisbury (60.9%), and Tea Tree Gully (56.3%), will all have a high rate of older population growth. Increased access to green and open spaces as well as greater affordability and a wider range of housing types are among the benefits of these LGAs, which will attract higher numbers of older people in the future. This is translated to lead to higher requirements for health and public service access, and therefore, measures need to be established or improved to enable the welfare of the older adults.

Adelaide Hills will have the second lowest rate of growth (20.9%) owing to its mountainous geography, which makes it difficult for older people to move around. Between 2016 and 2046, this is the only LGA in metropolitan Adelaide to witness population loss (−0.5%). This, combined with a scarcity of development potential and limited choices of dwelling types, explains population decline in the Adelaide Hills in recent years. Note that the projection method does not consider the structural characteristics of LGAs, such as upper limit and lower limit of carrying capacity, and intra-zonal migration of population, and each LGA is considered as an independent entity (Stillwell & Thomas, 2016). The average annual growth rate of older adults (1.9%) will be 2.5 times that of the general population, implying that more services and facilities will be required over the next three decades.

7.5.2 *Amenities Per Capita and Future Needs Based on Population Growth*

This section provides various categories of amenities per capita. For example, 98 POIs were identified inside Walkerville LGA boundaries. Community facilities (8), mobility (36), natural environs (5), recreation (15), services (12), and shopping (24) are all included. In 2016, the city had a population of 1662 older adults. As a result, the amenities per 100 persons are as follows: community facilities (0.48), mobility (2.17), natural environments (0.30), recreation (0.90), services (0.72), and shopping (1.44), respectively. We projected the amenities over 30 years using linear growth. It is anticipated that growth between 2016 and 2046 will follow the same trend as growth between 2016 and 2021 (Table 7.3).

Table 7.3 Amenities in 2016 and projection for 30 years by local government areas

LGA	Amenities _2016	Amenities _2016 (*10,000)	Amenity Growth (30 years)	Amenities _2046	Amenities _2046 (*10,000)	Amenities per capita_2016 (*1000)	Amenities per capita_2046 (*1000)
Adelaide	0.201	2007	0.21	2434	2149	63.8	77.4
Adelaide Hills	0.03	304	−0.243	0.023	230	2.8	2.1
Burnside	0.011	113	0.278	0.015	145	1.7	2.2
Campbelltown	0.015	151	0.625	0.025	246	1.8	2.9
Charles Sturt	0.008	83	0.434	0.012	120	1.1	1.6
Holdfast Bay	0.02	197	0.557	0.031	307	2.7	4.2
Marion	0.012	121	0.278	0.016	155	1.4	1.8
Mitcham	0.012	124	0.125	0.014	140	1.5	1.7
Norwood	0.022	220	0.029	0.023	227	3.3	3.4
Onkaparinga	0.008	78	0.192	0.009	92	0.9	1.1
Playford	0.011	112	0.254	0.014	141	1.6	2.0
Port Adelaide	0.012	119	−0.232	0.009	91	1.5	1.2
Prospect	0.029	286	0.111	0.032	318	5.4	6.0
Salisbury	0.011	108	0.558	0.017	169	1.1	1.7
TT Gully	0.012	115	0.153	0.013	133	1.2	1.4
Unley	0.026	261	0.088	0.028	284	3	3.3
Walkerville	0.022	218	0.55	0.034	339	3.7	5.7
West Torrens	0.013	132	0.087	0.014	143	1.2	1.3

*signifies figures in respective columns represent amenities per "x" number of old people

Figure 7.2 presents the cross-tabulation matrix between the amenity level and ageing level for the years 2016 and 2046. The results reveal that in majority of LGAs, access to facilities will remain low under the 'do-nothing' scenario. Campbelltown and Walkerville LGAs will see considerable improvements. The first will change from LY (low amenity, youngest old) to MY (medium amenity, youngest old), indicating the possibility of raising the amenity level. Walkerville will be upgraded from MY (medium amenity, youngest old) to HM (high amenity, middle old), demonstrating its continued achievement in providing increased amenity.

Some LGAs, including Charles Sturt, Holdfast Bay, Norwood-Payneham, Playford, Port Adelaide Enfield, Salisbury, and West Torrens, will see no substantial changes in age or amenity level. Adelaide and Prospect, on the other hand, will retain their top ranking in amenity level and younger cohort.

However, using an ANOVA, we discovered that there is no significant difference in amenities per capita between the LGAs ($F = 2.115$, $p < 0.155$). Because of its unique position, Adelaide was excluded from this test. However, the outcome of this test for population increase (2016–2046) demonstrates a statistically significant difference across the LGAs ($F = 10.794$, $p < 0.001$).

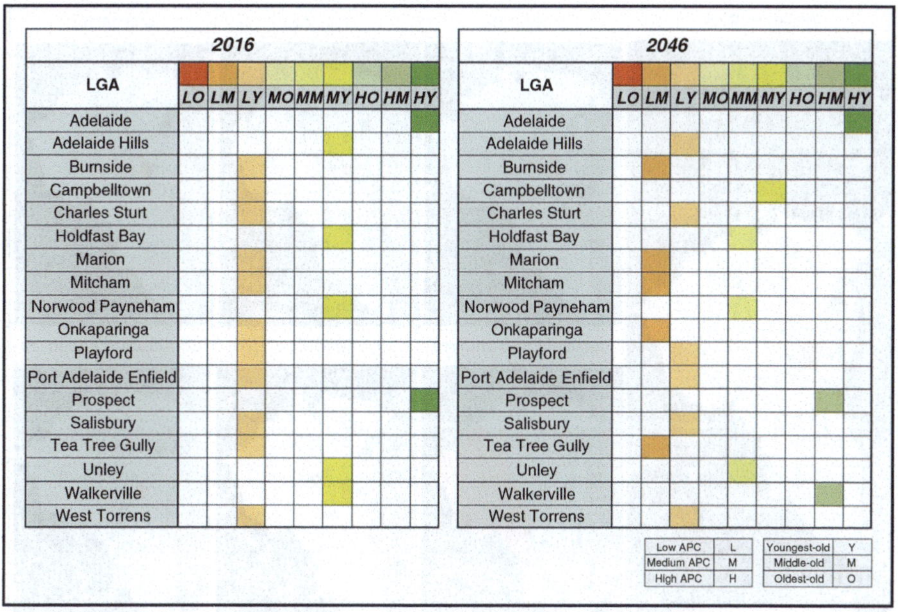

Fig. 7.2 Cross-tabulation matrix for amenity and age group. (Source: OpenStreetMap – Points of Interest (Australia) 2020; accessed from AURIN. https://data.aurin.org.au)

7.5.3 *Accessibility to Amenities*

As previously stated, we employed a gravity-based technique to evaluate the value of accessibility at the block level to each amenity. This method provides an opportunity to evaluate the performance of metropolitan Adelaide LGAs based on the provision of facilities for older adults. The results are presented in Fig. 7.3 based on access ranks. The findings are aggregated at the LGA level to allow for comparisons between LGAs.

The estimated average walking distance for each meshblock is utilised as the adaptive threshold to calculate the number of accessible amenities. Then, for each LGA, the cumulate accessibility is computed by averaging the accessibility values of all blocks within that LGA. The averaged accessibility estimates at the LGA level provide information on the level of accessibility to various facilities.

Figure 7.3 shows the six major categories derived from detailed land use activities. The spectrum from dark to light shades indicates decreasing levels of neighbourhood service accessibility. In overall, LGAs in the north and south have performed poorly in serving the requirements of the older adults, but LGAs in the central, western, and eastern central parts of the metropolitan area have performed better. However, there are substantial differences in the availability of various kinds of amenities among the LGAs. Furthermore, this study reveals that many sections of metropolitan Adelaide still lack essential amenities for senior residents. It is clear

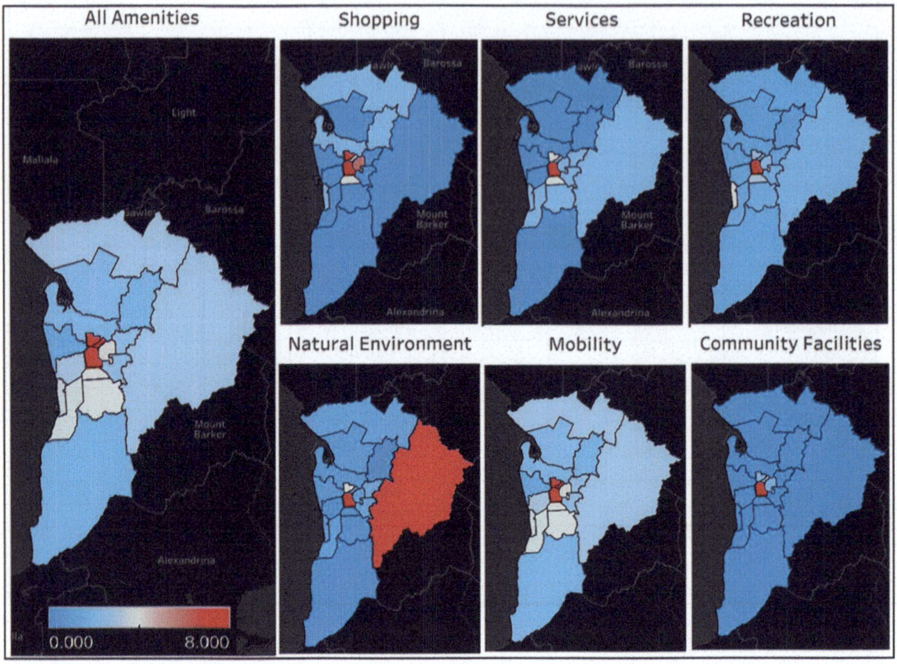

Fig. 7.3 Amenity types at LGA level. (Source: Census of Population and Housing 2016, Australian Bureau of Statistics; OpenStreetMap – Points of Interest (Australia) 2020; accessed from AURIN. https://data.aurin.org.au)

that the metropolitan area has the most categories of community facilities and services. The figure shows the number of parcels counted for each type of function, and it is noted that Adelaide LGA has the highest number of activities for shopping, followed by community facilities and services. These two types of activities are the foundational functions of nonresidential areas in the metropolitan area. Other functions, such as mobility and natural environments, contribute less.

Figure 7.3 shows empirical evidence of disparities in LGA-based walking accessibility to amenities among older individuals, showing significant differences in the level of accessibility. The results demonstrate that the walking access to amenities is spatially distributed, and the older adult group's residential choice of living patterns varies between LGAs. In general, older individuals may have more access to shopping than other services. Nonetheless, two-third live outside of the 400 m catchment area, which may influence older adults' quality of life (Wang et al., 2022). Older people have better access to services in core urban regions but have less access in the outer ring of the metropolitan area. Even though green and recreational spaces are more commonly situated in the outer suburbs (e.g. Adelaide Hills and Onkaparinga) than in the inner suburbs, this does not convert into better access levels due to the considerably longer walking distance to other amenities.

Because facilities are spatially separated and walking patterns to various tiers of services vary, variances in the amount of accessibility among LGAs are evident. In general, community amenities are more fairly distributed than other facilities. A notable concentration of all facilities is observed as spatial agglomeration in the Adelaide City area and several major inner suburbs to the north and east. Shopping locations are mostly concentrated in the southern and inner suburbs, while natural environment attractions tend to be more accessible in the eastern suburban regions and some outer suburbs in south.

Combining various types of services, centre, inner northern, and south-western districts are highlighted, while the eastern and northern outer suburbs are deficient, necessitating the demand for more built environment amenities. Furthermore, the distribution pattern of facilities leaves senior persons residing in the southern and northern outer suburbs with less fair access to total amenities.

Most amenities are located in the central and western areas. In the central urban area, uneven distribution of shopping, services, and community facilities is obvious, and recreation facilities are mainly concentrated in the inner suburbs. The amenities in the east are mostly concentrated around the major north-east routes. The same trend may be seen along the major south-west corridor. Most amenities such as shopping, and recreation have similar distribution patterns, with high concentrations in central and western areas, but other categories such as services have more equitable distribution patterns, with even distribution across metropolitan regions.

7.5.4 Accessibility to Amenities by Older Adults

It is found that the resident population covered by all amenities shows a significant downward trend with the increase of distance from their residence (Appendix 7.5). A small fraction of residents (65,426) lives within 400 m of an amenity, accounting for 32.3% of the total population. Within 400–800 m distance, the resident population reaches 55,497 people, accounting for 26.4% of the total residents. In other words, a significant proportion of older people (41.3%) are not located within the walkable distance threshold from the basic services.

7.6 Discussion and Conclusions

The aim of this research was to assess the accessibility of older adult cohorts to built environment facilities at the fine-grain neighbourhood scale in the Adelaide metropolitan area for supporting ageing in place. Comprehensively evaluating a built environment for ageing requires multiple methods including accessibility framework. To address the aim, an accessibility framework for older people to six key amenities were selected based on the literature (refer Table 7.1). We have compiled cohort of older adults throughout the metropolitan Adelaide to assess accessibility relationship of amenities. The information was collected on a meshblock scale, which implies that although having different individual characteristics, all people in the same meshblock had the same social and environmental characteristics.

This empirical research allowed us to conduct a comparative examination of location and accessibility. It made possible to evaluate the results at several sizes ranging from block to LGA levels. The study presented relationship between populations, location of amenities, and accessibility using GIS and data analytics methodologies for comparing service distribution among clusters of older persons. The data presented in the maps demonstrated how supportive or unsupportive the neighbourhood environment is for ageing in place.

The southern outer LGAs, for example, City of Onkaparinga, have low access index to amenities, whereas the City of Adelaide in the central area has the highest accessibility, indicating that locality has a significant impact on facilitating or limiting relatively capable access to the needs of older adults thus distribution equity (Logan et al., 2021). Only one fourth of older residents in Adelaide have access to facilities within 400 m of their homes. This is unfortunate since according to the WHO Checklist of Essential Features of Age-Friendly Cities, health and social services should be conveniently placed and accessible to older people via all modes of transportation. Our findings demonstrate that Adelaide metropolitan area is far from reaching age-friendly city requirements or the idea of the 15-minute city (Moreno et al., 2021; Duany & Steuteville, 2021).

We used demographic projection methods to forecast changes in the older adults over the next 30 years. The projection established that the growth rate of the older adults will be 2.5 times that of the general population. The forecast allowed us to identify the LGAs with a higher likelihood of attracting older citizens. The result demonstrated that besides central Adelaide, a number of inner eastern suburbs, such as Norwood and Payneham, will be more appealing to older adults in the future because of better access to health and public services. However, this might be limited to the high-income aged cohort because of relatively higher property prices in these LGAs. On the other hand, several LGAs in the outer north such as Playford or Onkaparinga in the outer south will absorb the older adult population because of low median property prices.

Table 7.4 presents categories of LGAs by priorities required to improve facilities to support ageing in place in 2046 for various LGAs based on the integration of three indexes: amenities per capita, future population growth, and ageing profile. Onkaparinga and Tea Tree Gully have high older ageing profile with lower level of facilities and are experiencing rapid population increase. Therefore, these LGAs should be given top priority to improve amenities to enable ageing in place. Whereas Charles Sturt, Playford, Port Adelaide Enfield, and Salisbury, which have low ageing profile and lower facilities provision and high population growth, should be the second priority to improve facilities and services in these LGAs. Burnside, Marion, and Mitcham should be the third priority LGAs to enhance amenities per capita, even though the ageing profile is higher because population growth and level of amenities are low. Other LGAs, except Adelaide Hills and West Torrens, are already

Table 7.4 Categories of LGAs by priorities required to improve facilities to support ageing in place in 2046

	Current Ageing profile	Amenities per capita_2046 (*1000)	Amenities per capita_2046 (*1000)	Population growth rate	Population growth rate	Priority for action
Onkaparinga	√√	1.1	√	1.166	√√	**1**
Tea Tree Gully	√√	1.4	√	0.955	√√	**1**
Playford	√	2.0	√	1.552	√√	**2**
Port Adelaide Enfield	√	1.2	√	1.029	√√	**2**
Salisbury	√	1.7	√	1.103	√√	**2**
Charles Sturt	√	1.6	√	0.959	√√	**2**
Burnside	√√	2.2	√	0.578	√	**3**
Marion	√√	1.8	√	0.819	√	**3**
Mitcham	√√	1.7	√	0.617	√	**3**
Unley	√√	3.3	√√	1.108	√	**4**
Walkerville	√√	5.7	√√	0.593	√	**4**
Norwood Payneham	√√	3.4	√√	0.745	√	**4**
Adelaide	√	7.1	√√	0.864	√	**4**
Campbelltown	√	**2.9**	√√	**0.818**	√	**4**
Adelaide Hills	√	2.1	√	0.588	√	**4**
West Torrens	√	1.3	√	0.465	√	**4**
Prospect	√√	6.0	√√	1.512	√√	**4**
Holdfast Bay	√√	4.2	√√	0.695	√	**4**

Note: √ = below Metro Adelaide; √√ = above Metro Adelaide
*signifies figures in respective columns represent amenities per "x" number of old people

with high per capita amenities. These two LGAs are in last categories because of low amenities per capita, low ageing profile, and low population growth.

The findings demonstrate that spatial distribution of amenities and accessibility to amenities have greater effect on older adult's choice of residence. However, proximity to an amenity is not always the only factor in ageing in place. Other factors such as individual circumstances, place attachment, safety, service quality, size of available facilities, access time, service frequency, usage cost, road congestion, network connectivity, parking provision, and service crowding are likely to influence the location choice for ageing in place (Vine et al., 2012).

The findings have far-reaching implications for both local governments and policymakers. With an increased emphasis on the ageing in place approach in Australia (O'Loughlin et al., 2017), empirical evidence of older adult's access to various health and public services is very useful in order to improve public service provision in South Australia and to overcome future challenges expected from increasing population ageing. One way to facilitate older adults' ability to remain independent is through the provision of basic required amenities (Cicero, 2012). Our study reflected that a large share (41%) of older adult's accessibility to daily amenities at the local level (catchments of 800 m) is not available. The study also reflected that public service facilities for older adults are predominantly concentrated in developed areas such as the central and western regions. Many important public amenities, on the other hand, are not available in the outer eastern and southern suburbs that have high concentration of older residents.

These could be achieved by mixed use zoning policy (Cheng et al., 2019). Moreover, the government and service providers need to modify their service allocation policies by determining the type of accessibility to health and public facilities and identifying geographical areas lacking such services (van Gaans & Dent, 2018). To achieve a balanced distribution of services that meet the actual needs of elderly individuals, it is necessary to go beyond the principles of supply and demand in the free market and price mechanism by introducing government subsidies (van Gaans & Dent, 2018). According to Talen (1997), the allocation of resources and facilities must be dispersed based on people' socioeconomic attributes. Furthermore, certain current basic services, particularly in the outer suburbs, might be improved to a more inclusive design.

The findings of this study show the extent to which ageing in place inequalities in access to health care and leisure activities exist in the metropolitan Adelaide. However, this study may be enhanced by increasing the database of public amenities and refining it to meet the requirements of older individuals (Dickins et al., 2020). The approach taken in this research can also be improved by including the diversity of characteristics of older adults (e.g. preferences, values, needs, and abilities), as well as their social structures and constraints that affect older adults' ability to translate resources and opportunities into practice. To effectively promote ageing in place, supporting services and amenities should be designed with the needs and preference of older population.

However, these initial findings show that the method has the potential to be generalised and modified in various contexts. This approach will help planners to examine the feasibility of ageing in place fits with the needs of older people. This will help stakeholders to figure out opportunities for understanding the gap and investments needed to enable ageing in place.

Review Questions

1. How would you define 'ageing in place'?
2. What are the factors that either facilitate or hinder the choice of ageing in place among older adults?
3. What are the reasons behind the increasing interest in the concept of ageing in place?
4. Why is the accessibility of older adults to built environment amenities at the fine-grain neighbourhood is important?
5. Which built environment amenities are essential for promoting ageing in place, and what are the reasons for their importance?
6. How can the method outlined in this chapter be used to analyse a neighbourhood and identify the fine-grain built environment amenities that either facilitate or impede ageing in place?

Acknowledgement We would like to thank the OpenStreetMap community for providing the data on points of interest (POI) used in this study. For more information, please visit the OpenStreetMap project website (https://www.openstreetmap.org/).

Appendices

Appendix 7.1: Mapping the Accessibility of Services at Two Radiuses of 400 and 800 m Based on Network Distance

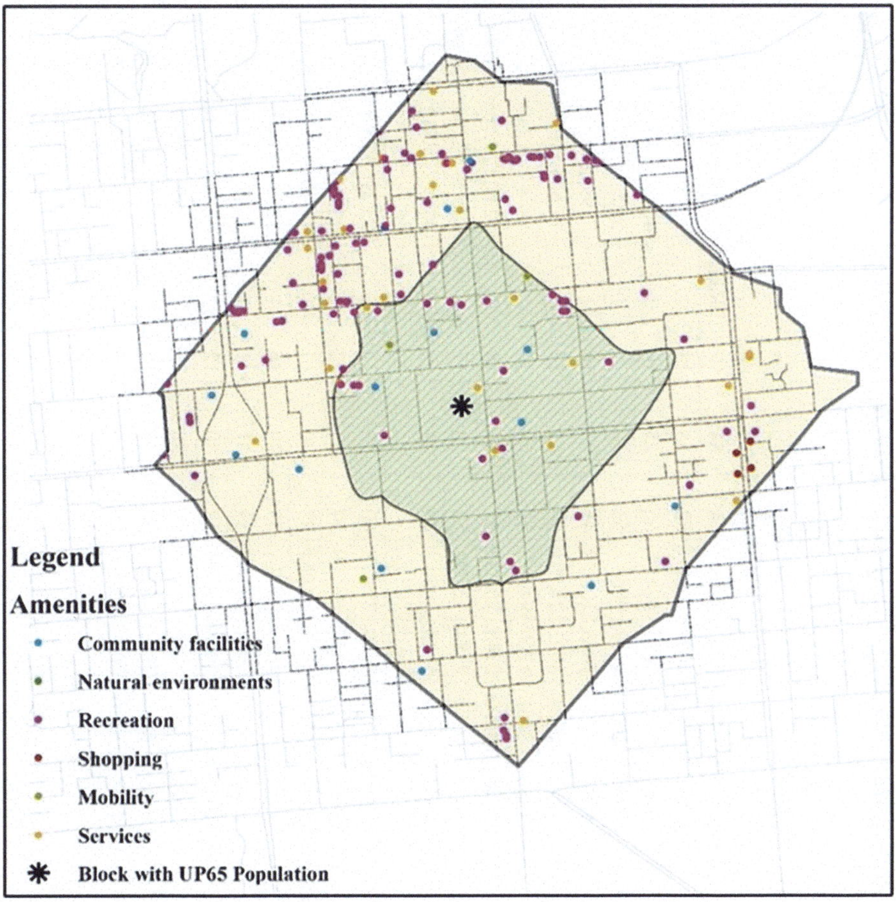

Source: OpenStreetMap – Points of Interest (Australia) 2020; accessed from AURIN. https://data.aurin.org.au

Appendix 7.2: Segmentation of LGAs Based on Age Profile

	2016								2046							
	65–74	75–84	85+	Total	65–74%	75–84%	85+ %	Range	65–74	75–84	85+	Total	65–74%	75–84%	85+ %	Range
Adelaide	1872	765	400	3037	62	25	13	1	3225	2384	1403	7013	46	34	20	1
Adelaide Hills	4368	1710	663	6741	65	25	10	1	4642	4051	2014	10,707	43	38	19	1
Burnside	5070	2861	1677	9608	53	30	17	1	5103	5870	4190	15,163	34	39	28	2
Campbelltown	4892	3797	1663	10,352	47	37	16	1	7199	6393	5232	18,824	38	34	28	1
Charles Sturt	10,675	6924	3692	21,291	50	33	17	1	17,605	16,796	7305	41,706	42	40	18	1
Holdfast Bay	4336	2461	1700	8497	51	29	20	1	5874	6230	2294	14,398	41	43	16	2
Marion	7888	4696	2748	15,332	51	31	18	1	12,027	12,086	3777	27,890	43	43	14	2
Mitcham	6629	3721	2184	12,534	53	30	17	1	8266	8296	3704	20,266	41	41	18	2
Norwood Payneham St Peters	3344	2277	1288	6909	48	33	19	1	4715	4812	2528	12,055	39	40	21	2
Onkaparinga	17,013	7906	3359	28,278	60	28	12	1	23,296	25,370	12,589	61,255	38	41	21	2
Playford	5859	3596	1335	10,790	54	33	12	1	14,147	9870	3515	27,532	51	36	13	1
Port Adelaide Enfield	10,058	5768	3138	18,964	53	30	17	1	17,893	15,369	5218	38,480	46	40	14	1
Prospect	1532	813	462	2807	55	29	16	1	2879	3119	1053	7051	41	44	15	2
Salisbury	11,030	5792	2018	18,840	59	31	11	1	16,679	15,107	7840	39,626	42	38	20	1
Tea Tree Gully	10,132	5156	2068	17,356	58	30	12	1	12,003	13,664	8261	33,928	35	40	24	2
Unley	3626	2009	1666	7301	50	28	23	1	4990	6802	3597	15,389	32	44	23	2
Walkerville	826	548	288	1662	50	33	17	1	1021	1036	591	2648	39	39	22	2
West Torrens	4648	3580	2068	10,296	45	35	20	1	6719	5430	2933	15,082	45	36	19	1

Appendix 7.3: Population Projection

The model for population projection of LGAs may have different forms such as below:

$$\text{Linear model} : x(t) = x_0 + rt \tag{7.2}$$

$$\text{Logarithmic} : x(t) = a * \log t + x_0 \tag{7.3}$$

$$\text{Polynomial} : x(t) = x_0 + at + bt^2 \tag{7.4}$$

$$\text{Exponential} : x(t) = x_0 e^{rt}, \log x(t) = \log x_0 + rt \tag{7.5}$$

$$\text{Logistic} : x(t) = \frac{K x_0 e^{rt}}{K - x_0 + x_0 e^{rt}}, 0 < x_0 \tag{7.6}$$

The used parameters are:

$x(t)$ = the size of a population at time t
x_0 = the size of a population at time 0
r = population growth rate (%)
t = time period for projection
a, b = equation coefficients
K = maximum population growth (the carrying capacity)

Appendix 7.4: Population Projection for 30 Years and Growth Rate by LGA

LGA	Population 2011[a]	Population 2016	Population 2021[b]	Equation	R^2	Population 2046	Aged 65+ 2046
Adelaide	19,638	23,552	27,617	$y = 3437.3x + 16619$	0.99	44,117	7013
Adelaide Hills	39,872	39,639	39,642	$y = -60.971x + 39848$	0.79	39,421	8827
Burnside	42,191	45,216	45,631	$y = 3148.8\ln(x) + 42442$	0.97	48,569	12,768
Campbelltown	49,649	51,079	53,059	$y = 47961e^{0.033x}$	0.99	62,700	16,139
Charles Sturt	108,456	114,977	121,110	$y = 5869.3x + 102995$	0.99	149,949	35,429
Holdfast Bay	36,763	36,532	37,230	$y = 77.018x^2 + 25.075x + 36480$	0.96	41,609	12,509
Marion	85,399	90,515	95,717	$y = 4464.3x + 81709$	0.98	117,423	24,486
Mitcham	65,057	66,486	67,119	$y = 1196x + 63802$	0.98	73,370	17,444
Norwood Payneham & St Peters	34,885	36,396	37,150	$y = 34225e^{0.027x}$	0.98	42,476	10,448
Onkaparinga	163,485	169,368	175,040	$y = 5584.2x + 158225$	0.99	202,899	47,801
Playford	80,748	90,549	100,356	$y = 9810.3x + 70846$	0.99	149,328	22,454
Port Adelaide Enfield	116,467	123,994	130,327	$y = 7263.9x + 109189$	0.99	167,300	32,585
Prospect	19,954	21,124	21,521	$y = 713.14x + 19434$	0.98	25,139	5278
Salisbury	132,499	140,370	144,694	$y = 13076\ln(x) + 131684$	0.98	158,875	30,322
Tea Tree Gully	98,028	99,144	101,319	$y = 183.93x^2 + 1421.9x + 96019$	0.98	116,407	27,124
Unley	36,844	38,939	38,792	$y = 1983\ln(x) + 36963$	0.90	41,087	11,561
Walkerville	7002	7781	7969	$y = 867.88\ln(x) + 7054.7$	0.98	8859	2294
West Torrens	56,960	59,413	60,944	$y = 54770e^{0.0378x}$	0.99	74,107	14,188
Metro Adelaide	1,134,387	1,191,883	1,237,978	NA	NA	1,563,637	338,668

[a]ABS (2022), [b]DTI (2021)

Appendix 7.5: Variation of Older Adult Populations Covered by Distance from Amenities

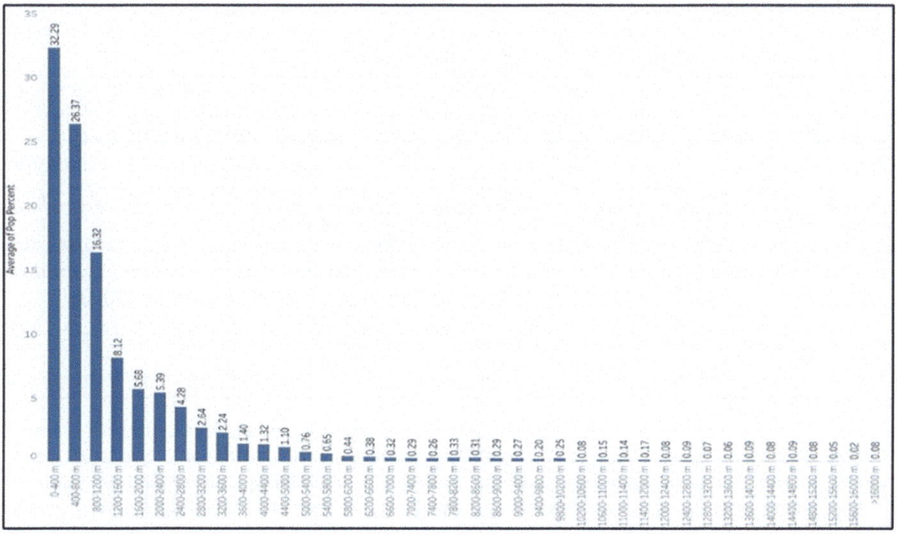

References

Abramsson, M., & Andersson, E. (2016). Changing preferences with ageing – Housing choices and housing plans of older people. *Housing, Theory and Society, 33*(2), 217–241. https://doi.org/10.1080/14036096.2015.1104385

ABS (Australian Bureau of Statistics). (2017). *Australian Demographic Statistics, June 2016.*

ABS (Australian Bureau of Statistics). (2022). Population: Census, Information on sex and age, reference period 2021. https://www.abs.gov.au/statistics/people/population/population-census/latest-release

Anselin, L. (1995). *Spatial econometrics: Methods and models* (Vol. 4). Springer Science & Business Media.

AURIN. (2021). *Walkability tools*. Australian Urban Research Infrastructure Network. https://docs.aurin.org.au/portal-help/analysing-your-data/walkability-tools/

Beard, J. R., & Petitot, C. (2010). Ageing and urbanization: Can cities be designed to foster active ageing? *Public Health Reviews, 32*(2), 427–450. https://doi.org/10.1007/BF03391610

BISA. (2018). *Submission for the South Australian Draft Public Health Plan 2019–2024*. https://www.bikeadelaide.org.au/wp-content/uploads/2018/06/201806-Submission-for-the-South-Australian-Draft-Public-Health-Plan-2019-2024.pdf

Bonvalet, C., & Ogg, J. (2008). The housing situation and residential strategies of older people in France. *Ageing and Society, 28*(6), 753–777. https://doi.org/10.1017/S0144686X08007071

Bowling, A. (2005). *Ageing well: Quality of life in old age*. Open University Press.

Brauer, F., & Castillo-Chavez, C. (2010). *Mathematical models in population biology and epidemiology*. Springer. https://doi.org/10.1007/978-1-4614-1686-9

Burton, E. J., Mitchell, L., & Stride, C. B. (2011). Good places for ageing in place: development of objective built environment measures for investigating links with older people's wellbeing. *BMC Public Health, 11*, 1–13. http://www.biomedcentral.com/1471-2458/11/839

Chan, E., & Lee, G. K. L. (2007). Critical factors for improving social sustainability of urban renewal projects. *Social Indicators Research, 85*(2), 243–256. https://doi.org/10.1007/s11205-007-9089-3

Chen, Y. (2015). A new methodology of spatial cross correlation analysis. *PLoS ONE, 10*(5), e0126158. https://doi.org/10.1371/journal.pone.0126158

Cheng, L., Caset, F., De Vos, J., Derudder, B., & Witlox, F. (2019). Investigating walking accessibility to recreational amenities for elderly people in Nanjing, China. *Transportation Research Part D: Transport and Environment, 76*, 85–99. https://doi.org/10.1016/j.trd.2019.09.019

Cheng, L., Yang, M., De Vos, J., & Witlox, F. (2020). Examining geographical accessibility to multi-tier hospital care services for the elderly: A focus on spatial equity. *Journal of Transport & Health, 19*, 100926. https://doi.org/10.1016/j.jth.2020.100926

Choi, K., Lee, Y., & Basrak, Z. (2021). Identifying communities of concern for older adults using spatial analysis: Focusing on accessibility to health, social, and daily services. *Journal of Applied Gerontology, 40*(11), 1527–1532. https://doi.org/10.1177/0733464820978000

Cicero, C. (2012). Ageing in place. In A. T. Carswell (Ed.), *The encyclopaedia of housing* (p. 928). SAGE.

Corbett, M., & Corbett, J. (2000). *Designing sustainable communities: Learning from village homes*. Island Press.

Cuthbert, A. R., & Dimitriou, H. T. (1992). Redeveloping the fifth quarter: A case study of redevelopment in Hong Kong. *Cities, 9*(3), 186–204. https://doi.org/10.1016/0264-2751(92)90015-W

Day, R. (2008). Local environments and older people's health: Dimensions from a comparative qualitative study in Scotland. *Health and Place, 14*(2), 299–312. https://doi.org/10.1016/j.healthplace.2007.07.001

de Boer, D., Hofstede, J. M., de Veer, A. J. E., Raijmakers, N. J. H., & Francke, A. L. (2017). Relatives' perceived quality of palliative care: Comparisons between care settings in which patients die. *BMC Palliative Care, 16*(1), 41. https://doi.org/10.1186/s12904-017-0224-x

de Jong, P. A., & Brouwer, A. E. (2012). Residential mobility of older adults in the Dutch housing market: Do individual characteristics and housing attributes have an effect on mobility? *European Spatial Research and Policy, 19*(1), 33–47. https://doi.org/10.2478/v10105-012-0004-9

Dempsey, S., Devine, M. T., Gillespie, T., Lyons, S., & Nolan, A. (2018). Coastal blue space and depression in older adults. *Health & Place, 54*, 110–117. https://doi.org/10.1016/j.healthplace.2018.09.002

Dickins, M., Joe, A., Enticott, J., Ogrin, R., & Lowthian, J. (2020). Trajectories of home nursing use for older women in Melbourne, Australia: 2006–2015. *Australasian Journal on Ageing, 39*(3), e295–e305. https://doi.org/10.1111/ajag.12735

Dolega, L., Pavlis, M., & Singleton, A. (2016). Estimating attractiveness, hierarchy and catchment area extents for a national set of retail centre agglomerations. *Journal of Retailing and Consumer Services, 28*, 78–90. https://doi.org/10.1016/j.jretconser.2015.08.013

DTI (Department for Trade and Investment). (2021). Population projections for South Australia and regions 2016–2041. https://plan.sa.gov.au/state_snapshot/populationold/population_projection/population_projections_for_south_australia_and_regions

Duany, A., & Steuteville, R. (2021). *Defining the 15-minute city*. DPZ CoDESIGN. Retrieved October 1, 2022, from https://www.cnu.org/publicsquare/2021/02/08/defining-15-minute-city

El-Geneidy, A., Grimsrud, M., Wasfi, R., Tétreault, P., & Surprenant-Legault, J. (2014). New evidence on walking distances to transit stops: Identifying redundancies and gaps using variable service areas. *Transportation, 41*(1), 193–210. https://doi.org/10.1007/s11116-013-9508-z

Eshelman, P. E., & Evans, G. W. (2002). Home again: Environmental predictors of place attachment and self–esteem for new retirement community residents. *Journal of Interior Design, 28*(1), 1–9. https://doi.org/10.1111/j.1939-1668.2002.tb00368.x

Finlay, J., Franke, T., McKay, H., & Sims-Gould, J. (2015). Therapeutic landscapes and wellbeing in later life: Impacts of blue and green spaces for older adults. *Health and Place, 34*, 97–106. https://doi.org/10.1016/j.healthplace.2015.05.001

Gan, D. R. Y. (2017). Neighborhood effects for ageing in place: A transdisciplinary framework toward health-promoting settings. *Housing and Society, 44*(1–2), 79–113. https://doi.org/10.1080/08882746.2017.1393283

Gariepy, G., Thombs, B., Kestens, Y., Kaufman, J. S., Blair, A., & Schmitz, N. (2015). The neighbourhood built environment and trajectories of depression symptom episodes in adults: A latent class growth analysis. *PLoS ONE, 10*(7), 1–13. https://doi.org/10.1371/journal.pone.0133603

Gehl, J. (1987). *Life between buildings: Using public space*. Van Nostrand Reinhold.

Gutiérrez, J. (2009). Transport and accessibility. In R. Kitchin & N. Thrift (Eds.), *International encyclopedia of human geography* (pp. 410–417). Elsevier. https://doi.org/10.1016/B978-008044910-4.01030-0

Hansen, W. G. (1959). How accessibility shapes land use. *Journal of the American Institute of Planners, 25*(2), 73–76. https://doi.org/10.1080/01944365908978307

Hansen, A., Bi, P., Nitschke, M., Pisaniello, D., Newbury, J., & Kitson, A. (2011). Perceptions of heat-susceptibility in older persons: Barriers to adaptation. *International Journal of Environmental Research and Public Health, 8*(12), 4714–4728. https://doi.org/10.3390/ijerph8124714

He, S., Li, C., Xiao, Y., & Liu, Q. (2022). Examining neighborhood effects on residents' daily activities in central Shanghai, China: Integrating "big data" and "thick data". *Environment and Planning B: Urban Analytics and City Science, 49*(7), 2011–2028. https://doi.org/10.1177/23998083221078307

James, A., Rowley, S., & Stone, W. (2020). *Effective downsizing options for older Australians*. Australian Housing and Urban Research Institute Report No. 325. https://doi.org/10.18408/ahuri-8118801

Jeger, M. J. (1984). The use of mathematical models in plant disease epidemiology. *Scientific Horticulture, 35*, 11–27. http://www.jstor.org/stable/45128399

Karuppannan, S., & Sivam, A. (2011). Social sustainability and neighbourhood design: An investigation of residents' satisfaction in Delhi. *Local Environment, 16*(9), 849–870. https://doi.org/10.1080/13549839.2011.607159

Lawton, M. P., & Nahemow, L. (1973). The psychology of adult development and aging. In C. Eisdorfer & M. P. Lawton (Eds.), *Ecology and the aging process* (pp. 619–674). American Psychological Association. https://doi.org/10.1037/10044-000

Lawton, M. P., & Simon, B. (1968). The ecology of social relationships in housing for the elderly. *The Gerontologist, 8*(2), 108–115. https://doi.org/10.1093/geront/8.2.108

Lecovich, E. (2014). Aging in place: From theory to practice. *Anthropological Notebooks, 20*(1), 21–32.

Logan, T. M., Anderson, M. J., Williams, T. G., & Conrow, L. (2021). Measuring inequalities in urban systems: An approach for evaluating the distribution of amenities and burdens. *Computers, Environment and Urban Systems, 86*, 101590. https://doi.org/10.1016/j.compenvurbsys.2020.101590

Lowen, T., Davern, M. T., Mavoa, S., & Brasher, K. (2015). Age-friendly cities and communities: Access to services for older people. *Australian Planner, 52*(4), 255–265. https://doi.org/10.1080/07293682.2015.1047874

Marsh, A. P., Miller, M. E., Saikin, A. M., Rejeski, W. J., Hu, N., Lauretani, F., Bandinelli, S., Guralnik, J. M., & Ferrucci, L. (2006). Lower extremity strength and power are associated with 400-meter walk time in older adults: The InCHIANTI study. *The Journals of Gerontology: Series A, Biological sciences and medical sciences, 61*(11), 1186–1193. https://doi.org/10.1093/gerona/61.11.1186

Moreno, C., Allam, Z., Chabaud, D., Gall, C., & Pratlong, F. (2021). Introducing the "15-Minute City": Sustainability, resilience and place identity in future post-pandemic cities. *Smart Cities, 4*(1), 93––111. https://www.mdpi.com/2624-6511/4/1/6

Mulliner, E., Riley, M., & Maiene, V. (2020). Older people's preferences for housing and environment characteristics. *Sustainability, 12*(14), 1–25. https://doi.org/10.3390/su12145723

Nath, T. K., Zhe Han, S. S., & Lechner, A. M. (2018). Urban green space and well-being in Kuala Lumpur, Malaysia. *Urban Forestry & Urban Greening, 36*, 34–41. https://doi.org/10.1016/j.ufug.2018.09.013

O'Loughlin, K., Browning, C., & Kendig, H. (Eds.). (2017). *Ageing in Australia: Challenges and opportunities* (Vol. 16). Springer.

Peace, S. M., Holland, C., & Kellaher, L. (2011). Option recognition in later life: Variations in ageing in place. *Ageing and Society, 31*(5), 734–757. https://doi.org/10.1017/S0144686X10001157

Pei, T., Sobolevsky, S., Ratti, C., Shaw, S.-L., Li, T., & Zhou, C. (2014). A new insight into land use classification based on aggregated mobile phone data. *International Journal of Geographical Information Science, 28*(9), 1988–2007. https://doi.org/10.1080/13658816.2014.913794

QUADRANT. (2022). *Point of Interest (POI) data as a service.* Retrieved August 2, 2022, from https://www.quadrant.io/

Robinson, D., Green, S., & Wilson, I. (2020). Housing options for older people in a reimagined housing system: A case study from England. *International Journal of Housing Policy, 20*(3), 344–366. https://doi.org/10.1080/19491247.2019.1644020

Rogers, W. A., Ramadhani, W. A., & Harris, M. T. (2020). Defining aging in place: The intersectionality of space, person, and time. *Innovation in Aging, 4*(4), igaa036. https://doi.org/10.1093/geroni/igaa036

Ryser, L., & Halseth, G. (2012). Resolving mobility constraints impacting rural seniors' access to regionalised services. *Journal of Ageing and & Social Policy, 24*(3), 328–344. https://doi.org/10.1080/08959420.2012.683329

Sala, G., Aboutaleb, A., Aki-Sawyerr, Y., Cantrell, L., Capp, S., Durkan, J., Kam-sing, W., Medina, F., Plante, V., Calle, D. Q., & Won-soon, P. (2020). *C40 mayors' agenda for a green and just recovery.* https://c40.my.salesforce.com/sfc/p/#36000001Enhz/a/1Q000000kVoY/kuR1PLHMGR2K9eEbo8aivV.xPegZVTqwt.EjX.4a.hk

Shrestha, R., Sivam, A., & Karuppannan, S. (2020). *Accessibility for seniors: Factors for improved accessibility to services*. Australian Sociological Association Conference.

Sivam, A., & Karuppannan, S. (2007). *Ageing and its implications for housing and urban development: South Australia.* State of Australian Cities (SOAC) Conference.

Sivam, A., & Karuppannan, S. (2008). *Factors influencing old age persons' residential satisfaction in Australia: A case study of South Australia.* Australian Sociological Association Conference, The University of Melbourne.

Soltani, A., Pojani, D., Askari, S., & Masoumi, H. E. (2018). Socio-demographic and built environment determinants of car use among older adults in Iran. *Journal of Transport Geography, 68*, 109–117. https://doi.org/10.1016/j.jtrangeo.2018.03.001

Stillwell, J., & Thomas, M. (2016). How far do internal migrants really move? Demonstrating a new method for the estimation of intra-zonal distance. *Regional Studies, Regional Science, 3*(1), 28–47. https://doi.org/10.1080/21681376.2015.1109473

Talen, E. (1997). The social equity of urban service distribution: An exploration of park access in Pueblo, Colorado, and Macon, Georgia. *Urban Geography, 18*(6), 521–541. https://doi.org/10.2747/0272-3638.18.6.521

Toole, J. L., Ulm, M., Bauer, D., & Gonzalez, M. C. (2012). Inferring land use from mobile phone activity. *arXiv, 1207.1115*. https://doi.org/10.1145/2346496.2346498

United Nations. (2020). *World population ageing 2020 highlights: Living arrangements of older persons.* Department of Economic and Social Affairs Population Division. https://www.un.org/development/desa/pd/sites/www.un.org.development.desa.pd/files/undesa_pd-2020_world_population_ageing_highlights.pdf

Van den Berg, A. E., Joye, Y., & Koole, S. L. (2016). Why viewing nature is more fascinating and restorative than viewing buildings: A closer look at perceived complexity. *Urban Forestry & Urban Greening, 20*, 397–401. https://doi.org/10.1016/j.ufug.2016.10.011

van Gaans, D., & Dent, E. (2018). Issues of accessibility to health services by older Australians: A review. *Public Health Reviews, 39*(1), 1–16. https://doi.org/10.1186/s40985-018-0097-4

Vecchio, G., & Martens, K. (2021). Accessibility and the capabilities approach: A review of the literature and proposal for conceptual advancements. *Transport Reviews, 41*(6), 833–854. https://doi.org/10.1080/01441647.2021.1931551

Veeroja, P. (2018). *The role of social and built environments in supporting older adults' social interaction* (Publication Number ORCID ID: 0000-0002-4994-4980). PhD thesis, The University of Melbourne. https://minerva-access.unimelb.edu.au/handle/11343/219679

Vine, D., Buys, L., & Aird, R. (2012). The use of amenities in high density neighbourhoods by older urban Australian residents. *Landscape and Urban Planning, 107*(2), 159–171. https://doi.org/10.1016/j.landurbplan.2012.05.013

Wang, S., Yung, E. H. K., & Sun, Y. (2022). Effects of open space accessibility and quality on older adults' visit: Planning towards equal right to the city. *Cities, 125*, 103611. https://doi.org/10.1016/j.cities.2022.103611

Wiles, J. L., Leibing, A., Guberman, N., Reeve, J., & Allen, R. E. S. (2012). The meaning of "aging in place" to older people. *The Gerontologist, 52*(3), 357–366. https://doi.org/10.1093/geront/gnr098

Wong, P.-H., Kourtit, K., & Nijkamp, P. (2021). The ideal neighbourhood of successful ageing: A machine learning approach. *Health and Place, 72*(1–13), Article 102704. https://doi.org/10.1016/j.healthplace.2021.102704

Zhang, J., Yu, Z., Zhao, B., Sun, R., & Vejre, H. (2020). Links between green space and public health: A bibliometric review of global research trends and future prospects from 1901 to 2019. *Environmental Research Letters, 15*(6), 063001. https://doi.org/10.1088/1748-9326/ab7f64

Zhang, X., Mu, L., & Shannon, J. (2021). The relationship between older adult migration and destination characteristics in Georgia. *Applied Geography, 132*, 102464. https://doi.org/10.1016/j.apgeog.2021.102464

Chapter 8
A Behavioural Approach to Sustainable Neighbourhoods: A Philosophical Construction of a Friendly Neighbourhood

Charles Prempeh

8.1 Introduction

Generally, sociologists admit that human beings are social beings whose sense of livelihood is aligned with their relations with others in society (Franke, 2021; Berger, 1967). Impliedly, human beings are embodied beings who live in a community where they expect to be accepted, away from all forms of multidimensional social exclusion. The need to build a human community of inclusivity necessitates the creation of a friendly neighbourhood where people's sense of worth as human beings is enforced. Meanwhile, the public sphere, expected to be rational and accommodating of pluralities (Habermas, 2005), where neighbourhood is formed is highly conflictual. Away from the immediate family environment, the public sphere is usually the space for economic and political activities, both of which are contentiously competitive. The contentious public sphere has become increasingly toxic for the construction of social conviviality. Issues of climate change, war/violence, infectious diseases, and industrialisation have reinforced the competition in the public sphere, as food production is on the decline, affecting health (Cobbina & Addaney, 2019). The recent coronavirus pandemic has also reinvigorated the globalisation of neoliberal capitalism, centring the West against the rest (Nhemachena et al., 2021). All this has burdened neighbourhood, resulting in the near collapse of social conviviality towards the aged.

Much as nearly everyone is affected by the decline in a friendly neighbourhood, which is indexed by issues such as criminality, violent attacks and different shades of security, the older generation is perhaps the most affected (Bempong & Asiamah, 2022). The reason for the skewed negative impact, burdening the older generation

C. Prempeh (✉)
Centre for Cultural and African Studies, Kwame Nkrumah University of Science and Technology, Kumasi, Ghana
e-mail: charles.prempeh@knust.edu.gh

N. Asiamah et al. (eds.), *Sustainable Neighbourhoods for Ageing in Place*,
https://doi.org/10.1007/978-3-031-41594-4_8

more, is because of the increasing trend in neoliberal economic logic. The increasing cost of living has reified society's appreciation for value for money more than value for life – accentuating Karl Marx's idea of commodity fetishism (Marx, 1990). Considering that the older generation hardly contributes to the chain of economic activities, including production and consumption, they are the most vulnerable to having their humanity questioned – indexing Orlando Patterson's idea of social death (Patterson, 1982).

As a solution, long-term residential care homes have increasingly become popular in several parts of the world, especially in the usually profiled developed West. Certainly, care homes – long-term care settings – have provided spaces for the older generation to find a place where they are professionally taken care of (Peace et al., 1997). Elsewhere in Ghana, the older generation is hemmed in communities called "witch camps" – pejoratively profiling and stigmatising them as witches (Adinkrah, 2015). The so-named witch camps are valorised as providing a safety net for old women against being hacked to death in their communities. Much as both care homes and witch camps have provided communities for the aged, the criticality of walkability, routinised in a friendly neighbourhood in advancing the health among the aged (Asiamah et al., 2023), is compromised. As I have said, neighbourhoods have increasingly become hostile and unsafe for the older generation. The issue, obviously, is to restore neighbourhood, indexed by improved access to transportation, healthcare facility, and community parks (Asiamah et al., 2023). These factors of building a friendly neighbourhood rely more on the technical answers (how answers) than the philosophical (why answers) to dealing with the multiple problems affecting the aged.

In my chapter, my goal is to focus on the nontechnical/philosophical aspect of building a friendly neighbourhood privileging more of the philosophy of what it means to be human than a technical response of safeguarding lives. For this purpose, the question informing my chapter is: Why is it that, despite the increasing growth and improvement in technology and healthcare practices and mushrooming care homes, the aged are often marginalised and neighbourhoods are hardly friendly? In response, my chapter argues that improvement in economic prosperity in the West, especially in America since the 1960s, and technology with the introduction of the Internet has rather produced a complex form of individualism. This individualism has increased in pace since the 1970s, nearly overtaking the capacity of society at recuperating the ethos of social conviviality that built human society. It is against this background of a near collapse social conviviality that Robert Bellah and Charles Taylor referred to the near collapse of community as "expressive individualism" and "age of authenticity", respectively (Bellah et al., 1985; Taylor, 2007).

In my reading, individualism centres on happiness, as an individual's ultimate goal as opposed to meaning, which is critical for building community (Baumeister et al., 2013). The quest for happiness has significantly morphed into individuals pursuing a career more than family – leading to a decrease in marriage and childbirth. At the same time, economic prosperity and improvement in health have together competed favourably against diseases – hence, fostering longevity (Xiong,

2021). At the same time, economic enrichment and technological advances produce their own nemesis with older men and women being left to fend for themselves or handed over to care homes and so-named witch camps. Meanwhile, care homes and witch camps hardly produce community for the older generation, who all their lives have found themselves surrounded by family and friends with whom they share faith and ancestry only to find themselves amidst "aliens" in their old age. As I have asked, why are the aged often relegated? Why is it that the increasing capitalist world cherishes value for money more than value for the sagacity of the sages? While appreciating varied responses to these questions, I answer these questions, as the basis of my chapter, from the philosophical prism of the ontological worth of the older generation.

Corroborating the technical answers to the creation of a friendly neighbourhood to support the aged, I push for the incorporation of Christian anthropology, which considers every human being as bearing the image of God as *conditio* sine qua non for reaffirming the ontological worth of the aged. By ontological worth, I argue that the older members of society are human beings regardless of their economic and technological contributions to society. It also means that they have inherent worth, rather than acquired worth that must be upheld as critical and necessary for establishing a friendly neighbourhood. Admittedly, however, my philosophical proposition is a highly idealistic answer to the challenges that have burdened the creation of a friendly neighbourhood for the aged. It is also limited to a particular religious community, Christianity with different strands that may hardly share my position. Nevertheless, given my own understanding of life as more about facing the reality to achieve the ideal, or filling in the gap between what is and what ought to be, my philosophical analysis of human ontological worth is my own contribution to pushing the bar away from the usual as part of advancing friendly neighbourhood. To contribute to routinising the philosophy of human ontological dignity in the socialisation of the younger generation, who have recently become more attuned to the social media world than the sagacity of the sages, I propose that the schools, particularly at the basic level, should incorporate religious anthropology of monogenesis (all human being emanating from one source) – in my particularistic case, the Christian idea of God creating human beings in His image – into the curriculum. I maintain that it is not enough just to teach the nineteenth-century evolutionary theory as the aetiological narrative for the origin of the human race.

The rest of the chapter is structured as follows: The next section provides my own knowledge and coming to a firm understanding of gerontocratic culture. This is followed by the impact of globalisation, at least limiting myself to the turn of the millennium on gerontocratic culture. With this background, I develop my theoretical framework, based on the idea of the human ontological dignity. The last two sections explore narratives about the family, notions of time, and the institution of care homes – all indexing a need for human ontological dignity to be considered critical. My conclusion brings out both the strengths and limitations of my theory.

8.2 Assessment of Gerontocratic Culture

Until recently at the turn of the millennium, several Ghanaians, mostly those born before the 1980s, tended to hold the impression that old age was sacred. Personally, as an early millennial born in 1982 and living in a Muslim-dominated inner-city of Maamobi, Accra, part of my socialisation is to cherish the aged. Through creative stories, myths, and legends, the older generation, and institutions, such as the *Madrasa* (Qur'anic schools) of the Muslims and Sunday Schools and Sabbath Schools of the Christians, my generation was told that there was a blessing associated with helping the aged. For the Muslims in particular, the aged were believed to hold *baraka* – blessings – that they graciously dispensed to youngsters who helped them. Refusing to undertake errands for the aged or cursing and maltreating them was considered a highly risky venture to undertake, as it brought curses that retards one's progress. As a highly multicultural community, indexed by the presence of multiple ethnic groups with usually shared traditions about gerontocratic culture, we were told that the aged had mystical power to bless and similarly withhold it.

My family, originally from the forest area of the Central Region, translocated the Akan mystification of gerontocratic culture to the city. They often told my siblings and me that "The mouth of the aged was more portend than the divine/deity". Certainly, this was to signal the perceived spiritual capacity that the older generation acquires as they go through life and reach its terminal end. Indeed, in an age-set society of so-called acephalous groups in Africa, the aged became ritual experts, providing ritual functionaries. This may be predicated on the assumptions that the aged have limited capacity to commit moral infractions, especially public delitescence that pollutes the soul. In effect, the older generation progresses to the point of "deities" with proximity to the ancestral world – affording them the position of intermediaries. For me, the idea that the aged had mystical power was brought home to me when at the age of 10, an elderly person performed a ritual that enabled a hearse van to function, when it came to a halt over a deceased resident of my community whose remains was brought to our community for final rites before onward burial in her village in the Central Region. Indeed, this was after the younger generation had run out of both their wits and physical energy in getting the mystically halted hearse van to recalibrate and continue its conveyance of the deceased from Accra to the village.

All the above come against the background of an imagined "Golden Age" for the aged, where the aged were treated with respect and their worth valued. Indeed, we could possibly argue that the sociopolitical structure of pre-European societies in African societies, particularly among the Akan, may have fostered young people's respect and deference to the aged (Sarpong, 1974) and that the aged controlled the means of production, particularly land and the materials for marriage, and determined prescribed marriages and also performed rituals that afforded them significant and enduring control over society (Awedoba, 2002). Nevertheless, far from romanticising such an imagined "Golden Age" for gerontocratic culture, the aged people in society have not always had a seamless and peaceful life, as they transitioned across to the other world. Especially during the colonial period when Western mercantilist logic prioritised transactional relationships over social conviviality, the aged had an unprecedented challenge. Their real or perceived economic incompetence structured their marginalisation in an increasingly labour-intensive colonial economy. The colonial establishment of plantation sites, the mushrooming of cities, and their interference with chieftaincy all weakened the authority of the aged (Frost, 1999). The introduction of Christianity and missionaries, who initially had less regard for chieftaincy and created the Salem residential segregated system in the case of the Basel missionaries in the early nineteenth century, rather created a State within a State (Boahen, 1975: 87). The new Christian converts felt compelled to obey their clerical leaders than the aged, who formed the base of the indigenous political system (Busia, 1951).

From colonial times through to postcolonial Ghana, the aged have had a difficult time getting integrated into society – getting worse as a result of neoliberalism (Kpessa-Whyte & Tsekpo, 2020; Geest, 1997; Apt, 1993; Apt, 1996). The continuation of colonial selective development – legitimating the charge that the African elites could not engineer freedom and the less impact of the decentralisation practice in ensuring even spread of social service – has sustained a regular flow of migration from the fringes to the cities (Awumbila & Ardayfio-Schandorf, 2008). All this has been complicated by the promotion of Western education, which – much as it has tipped to empower the educated – has created its own -emesis of graduates who tend to prefer white-collar jobs to farming and fishing. Ultimately, more of the younger generation are migrating into the cities or out of the country in search of greener pastures that usually are not readily available at their destinations. Those who are left behind in rural areas have also become part of the circuits of global consumption through the mediation of the ubiquity of social media and the Internet.

In the rural areas, the older generation are left to fend for themselves, while the females among them compound their multiple subsistence farming with caring for their grandchildren whose parents may have migrated to the cities (Frimpong-Nnuroh, 2004). I have observed, therefore, that unlike recent tension among French older members who are currently resisting the extension of the retirement age from 62 to 64 (The Associated Press, 2023), several of the aged constituency in Ghana and much of Africa hardly retire; they are retired from chronic diseases or eventually, death. Even in rural areas, the systemic collapse of gerontocratic culture that largely supported the age has exposed them to maltreatment from the younger generation (Apt, 1996).

8.3 Ageing in the Ghanaian Context

All these have been compounded by the metastasising impact of globalisation, indexed by climate change, war/violence, infectious diseases, and industrialisation. At the turn of the millennium, several academics discussed the possible impact of globalisation on the African or non-industrialised regions (Akokpari, 2000; Asante,

2001). Considered both an economic and political system that interlinked the world, the general understanding of several Africans was that globalisation would hardly inure to the benefit of Africans (Nyamnjoh, 2004). Instead, it was predicted to undermine the economic, cultural, and political independence of Africans and undermine the capacity of Africans to compete favourably in the world of competitive economic transactions and geopolitics (Nyamnjoh, 2004). The economic hardship that globalisation was predicted to engineer impacted outward migration that saw several Africans moving out of their respective countries, daring all foreseeable and unforeseeable challenges to reach their imagined "promised land" in Europe and the USA (Akokpari, 2000).

Meanwhile, since the beginning of the millennium, the improvement in science and technology, particularly artificial intelligence research, has created a functional differential where several human-induced labours had to be taken over by AI robots (Harari, 2016). This comes against the migratory narratives of the 1970s and 1980s, where the reconstruction of several post-World World European countries implied a need for human labour from Africa. As Europeans depend more on AI robots and also relocate several of their manufacturing industries to highly populated Asiatic countries, representing a major turn in neoliberal capitalism towards more consumption than production – jobs are not readily available for African to grab (Akurang-Parry, 2023). African migrants, including sometimes highly educated ones, are left to work in areas far below their high academic attainment (Akurang-Parry, 2023). At the same time, when extremist groups attacked the USA on 11 September 2001, the surveillance system progressed from just physical security checks to a form of a digital dictatorship. All this has resulted in stringent migration policies (Lazaridis & Wadia, 2015).

The impact of the above changing socioeconomic and geopolitics at both the local and global levels has had a predictable overwhelming impact on the aged – particularly the potentiality of a friendly neighbourhood. First, remittances have declined, while migrants tend to invest in economic enrichment programmes to generate profits rather than a wholesale project to support the aged. Second, the outflow of younger and more energetic people from the village have increased, leaving agriculture, since dependent on basic tools, in the hands of the aged. As I have said above, food security remains a challenge for some rural communities in Ghana.

In addition to the impact of globalisation on migration, the interconnectedness among nations has meant unequal carrying of burdens of the impact of global geopolitics. Largely informed by the idea of the "Anthropocene", the global West has sought to conquer nature, deconstructing myths that once preserved the ecosystem. In the end, the phenomenon of climate change is considered an existential reality. Climate change has been a much-discussed subject in both public and academic circles (Cobbina & Addaney, 2019), although it is highly controversial in terms of its impact and how to address it.

Meanwhile, various areas in rural Ghana continue to exhibit a decline in farming and fishing activities that are, inter alia, attributed to climate change. The extensive commodification of precious minerals, such as gold, and the desperation for it thereof have expanded illegal small-scale mining – known as *galamsey* (gather and sell) since the late 1980s (Afriyie et al., 2016). The impact of *galamsey* has been the young people's acquisition of land for mining, as opposed to farming, while water bodies are often polluted beyond their capacity to support flora and fauna. At the same time, the redirection of the focus of agriculture from producing crops to commercial crops for export also threatens food security among the Ghanaian popular in general and the aged in particular. Some of the proposals from the global West in the fight against corruption have resulted in environmental politics, where governments and other stakeholders are locked in a battle over whether to mine or not to mine in particular areas and what method to use. In the end, the inequality that has emerged from all this has increased corresponding insecurity, where the Government of Ghana's efforts at securitising land has rather created the unfortunate conflict between civilians and the security establishment – indexed by the grotesque murder of Major Maxwell Mahama at Denkyira-Obuasi in the Central Region in 2017. To be sure, if the life the security persons are hardly safe in any neighbourhood, then the older generation and their nexus with a friendly neighbourhood can at best be reflected in the Igbo proverb that "Old women feel uneasy when dry bones are mentioned in a proverb" (Achebe, 1989/1958).

8.4 Theoretical Framework

Considering the visible challenges that several aged people face in Ghana, it has become important to propose and apply a relatively new theory to gerontocratic studies. Considering the troubles that the aged go through in Ghana, including the multidimensionality of social exclusion, it could hardly be disputed that their humanity is questioned. Are older people considered human beings? This question may appear simplistic, trite, and easily answered. But it may not be that simple. For example, when we extend the question beyond just whether they are humans to ask about what makes one human, we enter into the domain of philosophical contestation that hardly has a uniformly acceptable response – at least, since the nineteenth-century Western Enlightenment. This question could also be asked: Do human beings have ontological dignity or inherent worth? This answer presumes an easy response, as several people ordinarily would not willingly want to harm another human being. But why would one deliberately harm another human being for any reason? If indeed, human beings have inherent worth, but not an acquired worth, why are older women either profiled as witches and consequently killed in some areas in Ghana or left to stay with "strangers" in care homes in the so-called developed world – even if some of these women have their cognitive and physical capacities intact to pass on virtues to their grandchildren and also walk around?

I reduce the above questions, informed by convenience, more than anything else to the Akan of Ghana. Far from being a homogenous group, the Akan generally share an anthropological perspective of human beings different from Christianity, introduced to them first in the fifteenth century and later in the nineteenth by European missionaries (Mohr, 2013). Among the Akan, human beings are not treated as ontological evil, in the same way Christian theology teaches concerning the fall of Adam and Eve into sin in the Garden of Eden (Quayesi-Amakye, 2013). Among the Akan, human beings have the capacity and willpower to both do good and evil. This is expressed in the many seemingly contradictory statements about human nature, such as "Human being is good, except that his or her meat is not eaten" and "Human beings have a heart full of wickedness". These conflictual statements are not to be read as a contradiction but as the Akan efforts at understanding human nature, which in the Akan philosophy is fluid and dynamic that cannot be reduced to a single ontological/dogmatic category.

In all of this, several efforts have been made by philosophers and social scientists to understand some of the excesses of human beings that promote *otherness* and *othering* for marginalisation. From the social evolutionary perspective of Friedreich Engels' origin of the family, the inequality in society was a result of the agricultural revolution (Engels, 1986). Until the agricultural revolution, society's natural disposition was towards equality, including men and women having free access to sex. But the agricultural revolution disrupted the flow and progressed to the industrial revolution era. The Industrial Revolution of the eighteenth century brought about seismic societal transformation for the Western human family. The major enduring impact of the progressive industrial revolution, merged with late capitalism is the measurement of human value based on economic worth, or what Karl Marx called "commodity fetishism" (Marx, 1990: 160).

This trend, which initially targeted the older generation, metastasised as a global impact, structured in multidimensionality of marginalisation through the lenses of social exclusion. The general Christian idea of inequality is mainstream through the dogma of the original sin, where disobedience of Adam and Eve created and retained a nosedive fall in interhuman and human-God relations. The Bible narratives take the story about such a downward spiral in human vertical and horizontal relationships from Genesis Chaps. 3, 4, 5, 6, 7, 8, 9, 10, 11, and 12, when God now experimented with the redemption of human beings through Abraham, an iconic figure whose piety is celebrated by the three Near Eastern Religions, Judaism, Christianity, and Islam.

Through the missionary activities of Christians in the first century and Muslims in the seventh century, both Christianity and Islam have become part of the religious or spiritual map of Africa (Mazrui, 1985). Considering that both Muslims and Christians have indigenised the religion on the patterns of inclusivity provided by indigenous cultures (Akrong & Azumah, 2013), all three religions tend to share a common narrative about the anthropological origin of human beings as a creature of God. While these traditions may diverge about the fine details of what it means for God to create human beings, whether in God's image, as in Christianity, or through the combination of elements, as is found in Islam, these religious traditions share the same general idea of creationism.

It is against the background of a shared general idea of creationism that my paper deploys Ali Mazrui's framing of Kwame Nkrumah's *consciencism* as Africa's triple heritage to discuss the idea of human ontological dignity (Nkrumah, 1964; Mazrui, 1986). The modern state, which is a product of the sixteenth-century Christian

Reformation, is a social and cultural construct (Anderson, 1991). From the perspective of social psychology, it is nearly impossible for an individual to establish an intimate bond with more than 150 people (Brooks, 2021). Thus, historically, nations have leveraged the mystification of the mundane to inspire nationhood (Hastings, 1997). The Asantes deployed the "myth" of the Golden Stool; the postcolonial state was built on both violence and legislation (Wilks, 1996). The issue was about how the postcolonial state was to be constructed. Nkrumah, Ghana's first president, was concerned about the existential religious differences and expansive multiculturalism that he and Ghanaians inherited from colonialism. He, therefore, used symbols, the mystification of himself, and rhetoric about colonialism to enforce a country of an imagined united people (Fuller, 2014).

His philosophy of *consciencism*, inter alia, implied the synthesisation of Ghana's religious plurality, mainly Islam, Christianity, and indigenous cultures, to foster national unity. Ali Mazrui picked the philosophy up in the 1980s and profiled it as Africa's triple heritage. The triple heritage remains an important philosophical reality that informs religious eclecticism among Ghanaians. It is my understanding of religious eclecticism, where Ghanaians consume rituals across religions in search of meaning, that I discuss human ontological dignity. By human ontological dignity, my point is that every human being has an inherent worth that must be protected. As a philosophical concept, it is based first on the religious community's acceptance of creationism as the primordial origin of human beings. But second, it continues from the Judeo-Christian idea that God created human beings in His image and likeness. This assertion has had different shades of interpretations from theologians. But as a non-theologian, I take that statement to mean that every human being belongs to God and none to do injustice to anyone human being, without God telling the perpetrator of injustice that God's creation has been violated.

From the above, I argue that human beings do not have to prove themselves to be accepted as human beings; they do not have to define their worth by their economic significance in contributing to productivity. Human beings are not dogs to be killed and gods to be offered cultic attention to exercise overwhelming control over others. From that, my own answer to the question of what it means to be human is simply that one is a human being independent of other human beings. This is not to deny the role society plays in offering individuals both legal and social identity – responsibilities and duties – as it is to say that every person has an inherent worth that must be respected, irrespective of all things. Socialisation helps human beings to attain social skills that foster conviviality; it does not make us human beings. It rather affirms the ontological dignity of a human being.

Relating this to a need to treat older people with respect, my point is that oftentimes, old people are the target of discontented young people, because members of the older generation are easily marginalised. They easily fall victim to the multidimensionality of social exclusion because their worth is measured in economic terms. As people who are also physically weak and sometimes with lifetime health challenges, including dementia, the older generation are easily disposed of supposedly saving family's income from being invested in their health. My theory, therefore, repudiates the young people's conflictual and condescending interaction with the older generation, which is usually routinised through social relation as an economic transaction. For this reason, my theory helps consider the capitalist adulation of money – which is always expressed tacitly as value for money but leaves no value for the aged.

8.5 Building Social-Oriented Neighbourhood

Sociologically, researchers have observed that human beings are social beings with a strong sense of gregariousness who find meaning through their interaction with others. Historically, therefore, different cultures have evolved several ways of accommodating people to foster neighbourliness. These creative approaches to building neighbourliness included drawing ethical boundaries that defined rights and responsibilities. It also included stories, myths, and legends that ensured that all, particularly the aged, are not relegated to the background. Among the Akan of Ghana, for example, a popular proverb had it that "All human beings are the children of God". By admitting that all human beings are the children of God, the Akan provided a standard against which everyone was to be accepted. This was consolidated in taboos that prohibited the disclosure of one's ancestry, including norms that ensured that even persons of alien and low social status could rise to occupy important social and political positions in society (Perbi, 2004). To be sure, this is not to create the impression that the Akan society was egalitarian in accommodating differences. For example, politically, the chiefs were more powerful, and by their role as mediators between the ancestors and their people in the physical world, the chiefs tended to control the spiritual map – which ordinarily was expected to be a democratised reality for all (Akyeampong & Obeng, 1995). Nevertheless, the recognition that every human being needed to be accepted was put into practice in such a way that communalism, however complex it was conceptualised, became very idealistic in advancing a friendly neighbourhood.

I have already mentioned above some of the structures that were put in place to foster inclusive neighbourliness in the demonstration of conviviality to the aged. I have similarly discussed the reasons, such as the rise in neoliberal economic practices that have significantly affected the aged. Meanwhile, there were yet other structures, such as consulting the aged that enhanced the life of the aged. Among the Akan, whenever the council of elders reached their wit's end in taking a decision, they deferred the matter to the aged. This was usually the case with genealogical issues where women, as primordial founders of the Akan, are expected to know the family's history. It is largely for this reason that the Akan said that it is a woman who births the chief. So, generally metaphorised as *Yɛkɔ hu abrewa* to wit: "We going to see the old woman", aged women were accorded respect in the neighbourhood (Ephirim-Donkor, 2021).

The aged as the storehouse and embodiment of genealogical knowledge had their position recognised in sociogenic activities. Generally, all the rites of passage among the Akan and several other African people are crystalised in a public celebration. For this reason, marriages, naming ceremonies, and funerals may go through channels of isolation of the neophyte (or the person going through the transition) and yet conclude with public social celebration. During the climaxing of such transitional stages in life, the aged move from their homes to perform rituals and also offer historical legitimation of the event. Such sociogenic activities, which involved walkability and mobility among the aged, created a social space for the aged to retain their continuing relevance in society.

Nevertheless, much as sociogenic activities still retain significance in the Ghanaian social space, the role of the aged rather is on the decline. The advent of the Internet in the 1990s and the social media revolution at the turn of the 2000s have broadened the channels of communication (Agana & Prempeh, 2022). Not only that, the channels of communication have included the privileging and centring of the voices of individuals who have acquired savviness in their use of technology more than their depth in the existential knowledge of life. I use existential knowledge to index epistemological traditions about life's questions such as follows: Who am I? Why am I here? Where do I go after here? These are questions that are usually answered and responded to through years of accumulation of experiences, other than just technological know-how. They are also not just about the rhetorical use of language, including foreign languages other than one's ancestral language. Impliedly, while existential knowledge is hardly static and open to re-interpretation, the sagacity of the sages can hardly be overlooked (Busia, 1962). Indeed, the younger generation relegates the voices of the aged on the "why" questions to their own peril – in terms of losing their identity and quest for meaning.

The complexity involved in the Internet drowning the voices of the aged is that technology tends to foster happiness more than meaningfulness. Meanwhile, happiness is more about the individual than the community. Happiness, which often tends to involve self-absorption, "self" exoneration over community is also about individualism, which social media often fosters. Social media tends to usually unite individual *self* to a common platform that tends to exclude dissenting voices. Through the formation of WhatsApp platforms and the structures of social media handles with icons that allow for more "like" than an expression of contrary opinions, social media tends to foster confirmation bias among individuals. The end is that social media does not ultimately satisfy the individual's quest for meaning (Turkle, 2011). Meanwhile, much as the quest for happiness is natural to nearly all living creatures, including animals, meaning is not. Meaning is a cultural creation of human beings that, as I have said, involves the extent to which individuals respond to life's existential questions. Answering the existential questions of life, if pursued as a goal, would significantly restore the voices of the aged to a world of technological fetishism.

My own observation is that the invention of social media has had its usefulness. As I have mentioned above, it has broadened the channels of communication and yielded the democratic benefit of allowing for the expression of alternative voices. But whether such attributable benefits of social media have ultimately helped in the democratisation quest of governance is not conclusive (Mutsvairo, 2016). Social media has also allowed individuals to form groups to advance their cultural and economic aspirations (Prempeh, 2021). But on the whole, social media has fostered multidimensional social marginalisation, which has significantly affected the creation of friendly neighbourliness for the aged. Meanwhile, meaning is essentially about a sense of purpose, which is indexed by one's significant decentring "self" to contribute to the good of society.

8.6 The Aged, Time, and Care Homes

My discussion has so far emphasised the mixed impact of the Internet and social media on the creation of a friendly neighbourhood to accommodate the aged. Nevertheless, as I have said, life is generally not about happiness, which tends to be present in focus. The social media world is also a world of "presentism", where an individual's quest for gratification is about "the now" of life. For this reason, as individuals pursue happiness, their notion of time is itself complexly affected by unilaterality. The past is considered needlessly "traditional" to avoid, while those who advocate a recuperation of the past could be treated as cultural atavists (Gyekye, 1998). At the same time, with social media creating an imagined world of happiness that is "now", the future is hardly considered.

Nevertheless, the Akan or, generally, human existence is not just about "today" as it is about the past and the future. The past provides the pattern and precedence for society to make sense of the present, even as they peep into the future. I am aware of John Mbiti's assertion that the African is more oriented to the present and the past than the future (Mbiti, 1969). That, in my understanding, may be construed to mean that Mbiti questioned the African sense of futurism. Personally, I do not think that was what Mbiti intended to communicate. The point is that the future is hardly known, and to sacrifice the present in the name of preserving an unknown has been the philosophical position of several African cultures. In other words, the African worldview is about here and now; there is no heaven for human beings to desire, which in some religions brings about human apathy towards making life better (Danquah, 1968). For this reason, the African emphasis on the present is not present that relegates the future as it is about the future that looks to a metaphysical blissful post-mortal space called "heaven" or "nirvana".

Again, limiting myself to the Akan, the idea that human beings dwell under the sun (*ɛwiase*) as their ontological home, the future – as in life away from the earth – is not part of their imagination. For this reason, the Akan worldview relies largely on what has happened before and what is happening now to make sense of what is yet to happen in the material world. Bridging time as cyclical, away from linear progression, provides a philosophical chronology that privileges the aged. It is against this background that, as I have said, the Akan privileged the voices of the aged, particularly women, in genealogical matters. Because the Akan, generally, is matrilineal, the woman's fecundity and cultural knowledge are cherished. Through fecundity and ancestral knowledge, old women played the role of performing fertility rituals, during rites such as *Bragoro* among the Akan and *Dipo* among the Krobo.

And through ancestral genealogical knowledge, Akan women's stool is treated as senior with anterior significance to her male counterparts (Rattray, 1927). I am aware of the debates over the position of the Akan *ɔhemaa*, which I deeply appreciate (Prempeh, 2023). Nevertheless, the point remains that the Akan woman retains an overwhelming political influence, both historically and contemporarily (Boni, 2001).

The centrality of the Akan woman, particularly the aged, has consequences for the family. Historically, the family, much as it is undergoing seismic changes towards nucleation, remains the basic unit of society. The state is highly imagined as the product of social and cultural construction, based on law, while the family, especially the mother-child bond among the Akan, is naturally constructed based on love (Prempeh, 2023). Also, the state is a social construct that manages competition in the public sphere of economic and political contestation to achieve happiness for an individual (Prempeh, 2023). On the other hand, the family is based on a social covenant, where the achievement of meaning is usually the focus (Prempeh, 2023). The family also thrives on the stories that are told. These stories, as I have said, are considered the preserve of the aged; more so, old women. It is for this reason that Akan women are major stakeholders in the socialisation of the younger generation, who pass on customs, values, and virtues, from their ancestors to the younger ones. This function allows new members of a family to understand their history and live meaningfully.

The near collapse of sagacious wisdom in the Western world and also its impact on several urban centres – indexed by the creation of aged homes – remain challenges. First, care homes limit the physical mobility of the aged. Certainly, some of the aged may have lost their mobility capacity, but being relocated from a familial space at home to a "strange" space, where a fictive family is created, rather makes life more difficult for them. Care homes have contributed significantly to "fossilising" the sublime wisdom of the sages, making the younger generation pursue an imagined "happiness" as opposed to meaning, based on the building of community. With the metastasising impact of globalisation, as I have discussed above, several of the aged constituency have been victimised by disillusioned young people.

In Ghana, the victimisation of older people has found expression in some individuals with prophetic missions instigating hatred against the aged. Usually, these persons with an interest in prophetic sermons deploy blame culture to stage older people as witches who are burdening the progress of the younger generation. To forestall, this trend, care homes have been very helpful.

Notwithstanding the benefits that the aged have derived from care homes, the issue is more complex than what such homes could resolve. First, the improvement in medicine and life-sustaining technologies would lead to a longer lifespan. Second, the increasing impact of late capitalism, which tends to pursue economic enrichment, as opposed to social relations, may affect the aged. In the end, while the world would be rich economically, the aged whose economic contribution may hardly be valued would suffer. In my estimation, therefore, the neoliberal world that fosters more value for money than value for life – more so the life of the aged – may cause the role of care homes to pale in significance. Quite difficult to say, the decline in the birth rate among the young population, as indexed in the case of China, portends a major challenge to the aged (Stevenson & Wang, 2023).

The complexities of the issues above imply that more than technology and care homes, much is needed to reinstate the dignity of the aged. As I have said, the aged are often not economically rated as important and are often side-lined. As retirees and sometimes nurturing illnesses, they may not have economic viability, while their contribution to the neoliberal world would not support the idea of value for money. But the question is, would the world consider the idea of value for life by privileging the lives of the aged? This question leads to my discussion on human ontological value as the best answer to fostering friendly neighbourliness for the aged.

8.7 Human Ontological Dignity and Friendly Neighbourhood

I have discussed the issues of time and the impact of the neoliberal economy on the aged. As I have identified the complexities of the ethos of the modern world, privileging happiness, and money, as opposed to community, burdens the lives of the aged. In this section, therefore, I discuss the issue of ontological dignity as a theoretical framework on how neighbourhoods can be re-designed or managed to support older people. For convenience reasons, I will focus more on a particular understanding of Christian anthropology. I must indicate that not every Christian will accept my interpretation of Biblical anthropology. For this reason, my views are what I deduce as a Christian and what I personally think could help revitalise the human community to accommodate the aged.

Largely, until Charles Darwin's publication of his magnum opus, *The origin of species*, in the nineteenth century provided a major alternative to Christian creationism, the Biblical aetiological creation story about human origin was largely accepted (Kuiper, 2010). But following Darwin's publication, Christian anthropology has come under sustained criticism; with some scholars generally considering it as a myth and unscientific. The voices of critics of Christian creationism have contributed to broadening the frontiers of knowledge. Nevertheless, given my discussion and a need for a solution, I personally think evolution theory, which has occasionally festered racism (Schuller, 2017), may not be of help. For this reason, I defer to Christian Anthropology.

From the perspective of Christian anthropology, God created man and woman in His image. This assertion has had different shades of interpretation, including the fact that human beings are rational and have a sense of creationism. But far more than that, I think the "image of God" on human beings is much more appreciated in light of human ontological dignity than anything else. This is because some animals may be more rational than human beings; while if we use the idea of rationalism and human creativity as the benchmark, we may relegate older men and women who may have lost their competence to reason according to the contemporary way of doing things or, as I have said, have lost their creative ability – in terms of their economic contributions.

Consequently, my understanding of God creating human beings in His image is that human beings have inherent value that cannot be negotiated under any circumstances. It means that human beings are cherished and appreciated regardless of their achievements or if they are absent of it. It also means that much as God offered human beings a divine-cultural mandate to rule the world, that cultural mandate is not just about economic enrichment – which having been overemphasised thrown the world into the Anthropocene stage (Williston, 2015). Instead, the cultural mandate includes the human capacity to formulate wisdom and philosophy to construct social conviviality, as God created human beings as dependent beings. As dependent beings, human beings are expected to foster a community that relies on the horizontal, fellow human beings, and vertical, God. This means that one must recognise God to recognise the other human being as ontologically human. Jesus restated the significance of both the vertical and horizontal relationships as the base of life in His summary of the law:

> Love the Lord your God with all your heart and with all your soul and with all your mind and with all your strength. The second is this: Love your neighbour as yourself. There is no commandment greater than these.[1]

In a world that is increasingly relegating the public role of religion, altruism and kindness towards all have been advocated as what would hold the impact of technology and individualism under check. More recently, I have personally interacted with people who sincerely advocate "love for your neighbour" to cure a world that they think is filled with more selfish people than ever. But sincere as these individuals want to see society taking a different turn to foster inclusiveness, focusing on just the horizontal without the vertical relationship, for me would only go far enough. After all, not much has been done to redirect education towards the promotion of philosophy. The complexity of secularisation and the overemphasis on scientism has significantly impacted funding for the humanities.

Meanwhile, in my own understanding, it is the humanities that, more than technology, better provide answers for the "why" questions. Equally true is that economics is more about individuals seeking their interest, which Adam Smith aptly captured as follows: "It is not from the benevolence of the butcher, the brewer, or the baker that we expect our dinner, but from their regard to their own self-interest" (Smith, 1981: 27). The point here is that to restore dignity to the aged, states and other stakeholders must, inter alia, focus on restoring Christian anthropology, as I have explained in terms of human ontological dignity, to the classrooms. This may help the younger generation to appreciate life in general and the aged in particular as, more than anything else, endowed by the Creator with an inherent worth, not acquired worth. This philosophical approach, as I understand it, will help in redesigning a neighbourhood that cherishes and respects the aged.

---

[1] Matthew 22: 37–38; Mark 12: 30:31; Luke 10: 27.

8.8 Implications for Ageing in a Sustainable Environment

That human beings have ontological worth that must be protected is not negotiable, as anything to the contrary would potentially foster the isolation of the aged and those considered economically and politically valueless for multidimensionality of social exclusion. It is, therefore, incumbent on governments and stakeholders of education to develop and roll out educational curriculum that reifies the philosophical foundation of the inherent worth of the aged. In easier way to safeguard this is to incorporate the *why* philosophical ethos of the various religious traditions in Ghana that foster respect for gerontocratic culture as an examinable subject in schools at the primary and secondary levels of education.

As I have indicated, the maiming and killing of older people, by younger people, usually on the slightest suspicion of an alleged older woman who is responsible for young people's predicament is a challenge to ageing. Nevertheless, it is very easy for the older generation to be the scapegoat for life's vicissitudes in the lives of the younger generation because of older people's vulnerability and inadequate enforcement of state's laws. Hence, policies aimed at protecting or supporting older people in a sustainable neighbourhood should be enforced by the government. There should be routine monitoring of the sermons of some neo-Pentecostal preachers, who tend to instigate violence against the older generation in the various regions of Ghana. All this also means that the state judiciary system should be resourced to expedite the trial of perpetrators of violence against the aged. Not just that, such violators of the rights of the aged should be seen punished to both serve as a caution to others who may desire to scapegoat the aged for their discontent with life.

8.9 Conclusion

My paper has brought out the complex challenges that have burdened the care offered to the aged. The metastasising impact of globalisation, indexed by the Internet revolution, social media, industrialisation, and climate change, and increasing neoliberal capitalist emphasis on value for money rather than value for life, a decline in population growth, increasing aged population, and an education system that fosters economic enrichment have had mixed impact on the aged. Economic and technological improvement has generated wealth, part of which has enabled the aged to receive care. Technology has fostered online communities and broadened the channels of communication, while skilled-oriented education appeared to have taken care of unemployment among the younger generation. But hardly accompanied by social skills, the increasingly aged population, whose economic contribution to the late capitalist economy is negligible, has rather rendered the aged more vulnerable. Care homes have provided some help. But given that care homes are public spaces, where care is given largely on an economic transactional basis, not everyone, particularly the impoverished, can afford it. In the end, the aged live in a neighbourhood – real or imagined – where their dignity is hardly respected. They are subjected to multidimensional social exclusion, including being profiled and consequently murdered. In my own reading and appreciation of the situation, the solution to the challenges of aged and ageing is the restoration of the philosophy of human ontological dignity as thought in the Bible. This means that human beings should be considered as, more than anything else, having inherent dignity, as opposed to an acquired dignity. This could be mainstreamed through the education system.

Nevertheless, my proposition is limited. It does not consider the position of other religious traditions, such as Islam. It also may not favour those who hold themselves as *religious nons*. So, going forward, future researchers and academics could explore other complementary sources of human dignity to foster the creation of enabling neighbourhoods to support the aged.

Review Questions

1. What does it mean to be a human being?
2. In what ways could religious anthropology about what it means to be human help provide sustainable environments for the aged?
3. Identify the shortfalls of ontological dignity in fostering respect for the aged in an increasingly industrialised world?
4. To what extent could the present generation draw from the past to consolidate the provision for inclusive friendly environments for the aged?

References

Achebe, C. (1989/1958). *Things fall apart*. Heinemann.

Adinkrah, M. (2015). *Witchcraft, witches, and violence in Ghana*. Berghahn.

Afriyie, K., Ganle, J. K., & Adomako, J. A. B. (2016). The good in evil: A discourse analysis of the *galamsey* industry in Ghana. *Oxford Development Studies, 44*, 493–508.

Agana, A. N., & Prempeh, C. (2022). Of farms, legends, and fools: Re-engaging Ghana's development narrative through social media. *Media, Culture & Society*, *44*(7), 1290–1306.

Akokpari, J. K. (2000). Globalisation and migration in Africa. *African Sociological Review, 4*(2), 72–92.

Akrong, A. A., & Azumah, A. (2013). Hermeneutical and theological resources in African traditional religions for Christian-Muslim relations in Africa. In J. Azumah & L. Sanneh (Eds.), *The African Christian and Islam* (pp. 65–84). Langham Monographs.

Akurang-Parry, K. (2023). *Velvet seekers: Africans orbiting these parts*. Digibooks Publishing.

Akyeampong, E., & Obeng, P. (1995). Spirituality, gender and power in Asante history. *The International Journal of African Historical Studies, 28*(3), 481–508.

Anderson, B. (1991). *Imagined communities: Reflections on the origin and spread of nationalism*. Verso.

Apt, N. A. (1993). Care for the elderly in Ghana: An emerging issue. *Journal of Cross-Cultural Gerontology, 8*, 301–312.

Apt, N. A. (1996). *Coping with age in a changing Africa: Social change and the elderly Ghanaian*. Avebury.

Asante, S. K. B. (2001). Africa and the challenges of globalisation: Agenda for action. *SAJEM, NS, 4*(3), 476–492.

Asiamah, N., Bateman, A., Hjorth, P., Khan, H. T. A., & Danquah, E. (2023). Socially active neighborhoods: Construct operationalization for aging in place, health promotion and psychometric testing. *Health Promotion International, 38*, 1–10.

Awedoba, A. K. (2002). *Culture and development in Africa: With special reference to Ghana*. Institute of African Studies.

Awumbila, M., & Ardayfio-Schandorf, E. (2008). Gendered poverty, migration and livelihood strategies of female porters in Accra, Ghana. *Norsk Geografisk Tidsskrift – Norwegian Journal of Geography, 62*(3), 171–179.

Baumeister, R. F., Vohs, K. D., Aaker, J., & Garbinsky, E. N. (2013). Some key differences between a happy life and a meaningful life. *Journal of Positive Psychology, 8*(6), 505–516.

Bellah, R. N., Madsen, R., Sullivan, W. M., Swidler, A., & Tipton, S. M. (1985). *Habits of the heart: Individualism and commitment in American life*. University of California Press.

Bempong, A. E., & Asiamah, N. (2022). Neighbourhood walkability as a moderator of the association between older Ghanaians' social activity, and the frequency of walking for transportation: A cross-sectional study with sensitivity analysis. *Archives of Gerontology and Geriatrics, 100*, 1–8.

Berger, P. (1967). *The sacred canopy: Elements of a sociological theory of religion*. Anchor Books.

Boahen, A. A. (1975). *Ghana: Evolution and change in the nineteenth and twentieth centuries*. Longman Group Limited.

Boni, S. (2001). Twentieth-century transformations in notions of gender, parenthood, and marriage in Southern Ghana: A critique of the hypothesis of 'retrograde steps' for Akan women. *History in Africa, 28*, 15–41.

Brooks, R. (2021). *Artificial intimacy: Virtual friends, digital lovers and algorithmic matchmakers*. Columbia University Press.

Busia, K. A. (1951). *The position of the chief in modern political system of Ashanti: A study of the influence of contemporary social change in Ashanti political institutions*. Oxford University Press.

Busia, K. A. (1962). *The challenge of Africa*. Pall Mall.

Cobbina, P. B., & Addaney, M. (Eds.). (2019). *The geography of climate change adaption in urban Africa*. Palgrave Macmillan.

Danquah, J. B. (1968). *The Akan doctrine of God: A fragment.of Gold Coast ethics and religion*. Routledge.

Engels, F. (1986). *The origin of the family, private poverty and the state*. Penguin.

Ephirim-Donkor, A. (2021). *African spirituality: On becoming ancestors* (3rd ed.). Hamilton Books.

Franke, N. (2021). *Built to belong: Discovering the power of community over competition*. Worthy Publishing.

Frimpong-Nnuroh, D. (2004). Ghost mothers and wet nurses: Breast feeding practices and care of children in Ellembelle Nzema. *Research Review, NS, 16*(2), 85–92.

Frost, D. (1999). *Work and community among West African migrant workers since the nineteenth century*. Liverpool University Press.

Fuller, H. (2014). *Building the Ghanaian nation-state: Kwame Nkrumah's symbolic nationalism*. Palgrave Macmillan.

Geest, S. V. D. (1997). Money and respect: The changing value of old age in rural Ghana. *Africa, 67*(4), 534–559.

Gyekye, K. (1998). Person and community in African thought. In P. H. Coetzee & A. P. J. Roux (Eds.), *Philosophy from Africa* (pp. 317–336). ITP.

Habermas, J. (2005). *Democracy and the public sphere*. Pluto Press.

Harari, Y. N. (2016). *Homo deus: A brief history of tomorrow*. Harvill Secker.
Hastings, A. (1997). *The construction of nationhood: Ethnicity, religion and nationalism*. Cambridge University Press.
Kpessa-Whyte, M., & Tsekpo, K. (2020). Lived experiences of the elderly in Ghana: Analysis of ageing policies and options for reform. *Journal of Cross-Cultural Gerontology, 35*, 342–352.
Kuiper, K. (Ed.). (2010). *The Britannica guide to theories and ideas that changed the modern world*. Britannica Educational Publishing.
Lazaridis, G., & Wadia, K. (Eds.). (2015). *The securitisation of migration in the EU: Debates since 9/11*. Palgrave Macmillan.
Marx, K. (1990). *Capital: A critique of political economy* Vol. I (trans. Ben Fowks and intro. By Ernest Mandel). London: Penguin Classics, 160.
Mazrui, A. A. (1985). Religion and political culture in Africa. *Journal of the American Academy of Religion, 53*(4), 817–839.
Mazrui, A. A. (1986). *The Africans: A triple heritage*. BBC Publications.
Mbiti, J. S. (1969). *African religions and philosophy*. Heinemann.
Mohr, A. (2013). *Enchanted Calvinism: Labor migration, afflicting spirits, and Christian therapy in the Presbyterian Church of Ghana*. University of Rochester Press.
Mutsvairo, B. (Ed.). (2016). *Digital activism in the social media era: Critical reflections on emerging trends in sub-Saharan Africa*. Palgrave Macmillan.
Nhemachena, A., Makamani, R., & Mtapuri, O. (2021). Natures' role in 21st century revolutions? Covid-19, the fourth revolution and reposition of global capital. In R. Makamani, A. Nhemachena, & O. Mtapuri (Eds.), *Global capitalist's 21st century repositioning: Between Covid-19 and the fourth industrial revolution on Africa* (pp. 1–18). Langaa Research & Publishing CIG.
Nkrumah, K. (1964). *Consciencism: Philosophy and ideology for decolonization and development*. Heinemann.
Nyamnjoh, F. B. (2004). Globalisation, boundaries and livelihoods: Perspectives on Africa. *Identity, Culture and Politics, 5*(1&2), 37–59.
Patterson, O. (1982). *Slavery and social death: A comparative study*. Harvard University Press.
Peace, S. M., Kellaher, L., & Willocks, D. (1997). *Re-evaluating residential care*. Open University Press.
Perbi, A. A. (2004). *A history of indigenous slavery in Ghana: From the 15th to the 19th century*. Sub-Saharan Publishers.
Prempeh, C. (2021). From offline to online imagined community: Recuperating Asante culture and history for development in Ghana. *Question, 06*, 36–44.
Prempeh, C. (2023). *Gender, sexuality and decolonisation in postcolonial Ghana: A socio-philosophical engagement*. Langaa RPCIG.
Quayesi-Amakye, J. (2013). *Christology and evil in Ghana: Towards a Pentecostal public theology*. Rodopi.
Rattray, R. S. (1927). *Religion and art in Ashanti*. Oxford University Press.
Sarpong, P. (1974). *Ghana in retrospect: Some aspects of Ghanaian culture*. Ghana Publishing Corporation.
Schuller, K. (2017). *The biopolitics of feeling: Race, sex, and sex in the nineteenth century*. Duke University Press.
Smith, A. (1981). *An enquiry into the nature and causes of the wealth of nations* (Vol. I). LIbertyClassics.
Stevenson, A., & Wang, Z. (2023, January 22). *China needs couples to have more babies: Can I.V.F. help?* https://www.nytimes.com/2023/01/22/business/china-birthrate-ivf.html?action=click&pgtype=Article&state=default&module=styln-china&variant=show&region=MAIN_CONTENT_1&block=storyline_top_links_recirc.
Taylor, C. (2007). *A secular age*. The Belknap Press of Harvard University Press.

The Associated Press. (2023, March 16). *Macron puts his government at risk with a decision to raise the retirement age in France*. https://www.npr.org/2023/03/16/1163863267/france-retirement-pension-strikes-protests-vote.

Turkle, S. (2011). *Alone together: Why we expect more from technology and less from each other*. Basic Books.

Wilks, I. (1996). *One nation, many histories: Ghana past and present*. Ghana Universities Press.

Williston, B. (2015). *The Anthropocene project: Virtue in the age of climate change*. Oxford University Press.

Xiong, X. (2021). Bring technology home and stay home: The role of fourth industrial revolution and technology in improving the efficacy of health care spending. *Technology Forecasting and Social Change, 165*, 1–6.

Chapter 9
Assistive Technologies for Ageing in Place: A Theoretical Proposition of Human Development Postulates

Nestor Asiamah, Emelia Danquah, Sarra Sghaier, Henry Kofi Mensah, and Kyriakos Kouveliotis

Acronyms

ADLs	Activities of daily living
ATA	Activity theory of ageing
ATs	Assistive technologies
CTA	Continuity theory of ageing
DTA	Disengagement theory of ageing
E-UTAUT	Extended unified theory of acceptance and use of technology

N. Asiamah (✉)
Division of Interdisciplinary Research and Practice, School of Health and Social Care, University of Essex, Colchester, Essex, UK

Department of Gerontology and Geriatrics, Africa Centre for Epidemiology, Accra, Ghana
e-mail: n.asiamah@essex.ac.uk; nestor.asiamah@ace-gh.org

E. Danquah
Department of Gerontology and Geriatrics, Africa Centre for Epidemiology, Accra, Ghana

Koforidua Technical University, Koforidua, Ghana
e-mail: addom123@yahoo.com

S. Sghaier
Department of Gerontology and Geriatrics, Africa Centre for Epidemiology, Accra, Ghana
e-mail: sarra.sg2@gmail.com

H. K. Mensah
Department of Organizational and Human Resources Development, Kwame Nkrumah University of Science and Technology, Kumasi, Ghana
e-mail: hkmensah@knust.edu.gh

K. Kouveliotis
Department of Gerontology and Geriatrics, Africa Centre for Epidemiology, Accra, Ghana

Office of the Provost, Berlin School of Business and Innovation, Berlin, Germany
e-mail: kyriakos.kouveliotis@berlinsbi.com

N. Asiamah et al. (eds.), *Sustainable Neighbourhoods for Ageing in Place*,
https://doi.org/10.1007/978-3-031-41594-4_9

IADLs	Instrumental activities of daily living
JD-R	Job demands-resources
MeSH	Medical subject headings
SCT	Social cognitive theory
SN	Serial number
SPIDER	Sample, phenomenon of interest, design, evaluation, and research type
SST	Socioemotional selectivity theory

9.1 Introduction

Population ageing implies that the proportion of older adults (e.g. people aged 65 years or higher) will grow over time (da Silva Francisco, 2017; Menassa et al., 2023). A decline in physical and cognitive functional ability due to a gradual loss of bone and muscle mass (i.e. sarcopenia) and the weakening of bones (i.e. osteoporosis) are the consequence of ageing (McPhee et al., 2016; Sezgin et al., 2019). Older adults with these age-related physiological conditions face a higher risk of frailty and physical disability, and many of them would be unable to perform activities of daily living (ADLs) and instrumental activities of daily living (IADLs) independently in later life. ADLs are self-care activities, such as eating, bathing, and toileting performed by the individual daily, whereas IADLs include social activities (e.g. visiting a grocery shop) by which individuals live independently in their community (Mois & Beer, 2020; Takatori et al., 2018). The maintenance of ADLs and IADLs is necessary for quality of life. Yet, the population of older adults without the ability to perform ADLs and IADLs is expected to increase as ageing intensifies.

People with physical and cognitive limitations often need assistance or support to perform ADLs and IADLs. Research to date shows that assistive technologies serve as social or physical activity aids for these individuals or are effective at enabling them to perform ADLs/IADLs (Claes et al., 2015; Mois & Beer, 2020). Assistive technology is any product, tool, service, or strategy used by older adults or people with disability to prevent or alleviate a disability or disadvantage and enhance autonomy as well as the quality of life (Pedrozo Campos Antunes et al., 2019). As this definition suggests, the literature predominantly views assistive technologies as equipment for facilitating ADLs/IADLs and health-seeking behaviours among people with disabilities. Some researchers (Dratsiou et al., 2021; Sen et al., 2022), though, have observed that assistive technologies can be used by everyone to maintain health and engagement with life (i.e. participating in social and economic activities such as employment). We adopt this view in this chapter since it supports the interest of stakeholders to prevent disease and maintain population health over the life course. We, therefore, define assistive technologies as products, tools, services, or strategies used by people to maintain health and engagement with life across the life span. With this definition, we suggest that individuals should be supported to retain their ability to use assistive technologies in later life.

Four global crises that can adversely affect ageing in place are climate change events, infectious disease outbreaks, extreme industrialisation (i.e. negative industrial practices), and violence. Infectious diseases can increase the risk of mortality and multimorbidity (Barchielli et al., 2022). Injuries and disabilities from war and climate change events (e.g. flooding, and hurricanes) are a public health emergency (Brockie & Miller, 2017; Gamble et al., 2013) as they can increase morbidity and mortality. Hazards (e.g. gas explosions, spillages, and pollution) from negative industrial practices have also caused fear and panic in communities (Asiamah, 2022; Ntengwe, 2006) and may compel residents to leave their neighbourhoods and homes. Researchers (Asiamah, 2022; Brockie & Miller, 2017) are concerned that older adults with physical and cognitive impairments are the most vulnerable to these crises, and their vulnerability is expected to increase as the frequency and intensity of the crises increase over time. The crises (e.g. floods and storms) can also damage assistive technologies or cut the supply of utilities or services (e.g. Internet and electricity) on which they depend, which has consequences for ageing in place.

Some studies (Becker et al., 2020; Khasnabis et al., 2020; Mois & Beer, 2020) suggest that assistive technologies can facilitate ageing in place, which is defined as the maintenance of health and engagement with life in one's home and neighbourhood in later life (Pani-Harreman et al., 2021). Ageing in place includes a sense of attachment to one's neighbourhood (Carnemolla, 2018; Pani-Harreman et al., 2021) and the maintenance of ADLs/IADLs. Even so, there are barriers to ageing in place, and some of the most pronounced ones are older adults' unwillingness and lack of ability to use assistive technologies (Pruchno, 2019; Yusif et al., 2016), which has crucial human development implications. Yet, there is a paucity of evidence providing an understanding of these implications for ageing in place in contexts experiencing crises. There is also no theoretical view to human development programmes aimed at enabling ageing people to maintain the effective use of assistive technologies across their life span, not only in later life when they are with disabilities.

In this chapter, therefore, we employ a narrative review to synthesise relevant theories and analyse the human development implications of older adults lacking or losing the ability to use assistive technologies sustainably, especially amidst crises. We relate the implications with the *technology demands-resources concept*, which embodies three individual states. This chapter comprises seven sections, with the first introductory section serving as a background to the chapter. In the second section, we present a summary of our review methodology and key findings on how assistive technologies can support people to age in place during crises. The theoretical framework is formulated in the third section, whereas the technology demands-resources concept is introduced in the fourth section. In the fifth section, we propose human development postulates to guide future interventions. Limitations and future research directions are reported in the sixth section, whereas the seventh section presents concluding remarks.

9.2 Review Methodology and Key Findings

This section presents a summary of the literature search methodology and findings from the search regarding different ways assistive technologies have been used by older adults or for ageing in place.

9.2.1 Search Methodology

Our search aimed to identify published documents on how assistive or information technologies have been used for ageing in place or by older adults. We performed the first search in the first week of December 2022, whereas the second one was performed in the first week of January 2023. The databases searched were Scopus, PubMed, PsycINFO, CINAHL, and Google Scholar. The search was based on primary key terms (i.e. "assistive technologies", ageing, "ageing in place", "older adults", health, environment, "human development") reached with the SPIDER (Sample, Phenomenon of Interest, Design, Evaluation, and Research Type) framework, which is suitable for reviews incorporating both qualitative and quantitative studies (Methley et al., 2014). Synonyms of the key terms were identified using MeSH (Medical Subject Headings) terms on PubMed. The Boolean operator "OR" was used to separate synonyms or related terms to generate sub-strings, which were subsequently combined with "AND". As an alternative to quality assessment, we focused on documents published by reputable organisations such as the United Nations or journals indexed by Scopus. We excluded all documents not written in English. A total of 1121 documents including 502 duplicates were downloaded. After applying the inclusion criteria, 32 documents were deemed appropriate. Two of the authors independently screened for the eligible documents after removing duplicates and resolving minor disagreements in the screening process.

9.2.2 Assistive Technologies and Ageing in Place

Ageing in place depends on whether the individual can avoid disease and a major decline in cognitive and physical functional capacity in the ageing process, which is why assistive technologies designed to support people with disabilities have become an important part of resources for ageing. Today, older adults with health problems requiring permanent admission into a care home can continue living in their homes with assistive technologies (Dratsiou et al., 2021; Hampel & Kunze, 2023). Though such individuals may not have optimal health, they may prefer living at home with technological aids to residing in a care home. Many people admitted into institutional care may not have carers (e.g. health professionals and relatives) who can commit to their daily care, but it is now possible for assistive technologies to fill the roles of these carers, enabling people with disabilities to remain at home in later life. This notwithstanding, we argue that a "life course approach" to the use of assistive technologies is more promising in a world facing rapid population ageing. This approach is about enabling people to effectively use assistive technologies to maintain health across the life span. It is superior to supporting people to use assistive technologies later in life when they are with health problems as it can curb a potential increase in disease burden and healthcare expenditure accompanied by population ageing.

Table 9.1 shows six categories of different ways assistive technologies have been used to support or facilitate ageing in place. It can be observed that all the methods can be applied to influence health-seeking application of assistive technologies across the life span. The first category comprises homecare, remote clinical care by health professionals (excluding physicians), and telemedicine and physician support. A noteworthy category that may play a critical role during crises is the sixth category (i.e. safety and security). Since the crises generally cause safety concerns (Asiamah, 2022; Barchielli et al., 2022), assistive technologies may be necessary for monitoring one's home, health, and wider physical environment to identify threats to safety and security. Being able to use assistive technologies to secure one's home and wider physical environment as well as health can curb long-term or permanent admission into a nursing home. Moreover, all the methods in the table are applicable to the general population, which means they support our proposed life course approach.

In contexts experiencing the crises, though, the possibility of assistive technologies supporting ageing in place and the life course approach may be very low or nonexistent. A climate change event, such as flooding, can cut the electric power supply or Internet connectivity that an assistive technology requires to function. Injuries and anxiety linked to this event may deprive older adults (especially those with physical and cognitive impairments) of the skills they need to continue using assistive technologies. With the following theoretical framework, we explain why ageing people must be supported to maintain the ability to use assistive technologies across the life span.

9.3 Theoretical Framework

The role of assistive technologies in ageing in place is implied by several theories, one of which is the disengagement theory of ageing (DTA) developed in the early 1960s (Asiamah, 2017). This theory asserts that the individual experiences a gradual decline in resources (e.g. social networks and income) and physical and cognitive ability over the life course. Consequently, health-supporting behaviours such as the use of assistive technologies for health can reduce in the ageing process. Individuals often need income (through employment) and human facilitators (e.g. a caregiver) to sustainably use assistive technologies (Peek et al., 2014), which implies that ageing people lose the ability to use assistive technologies across the

Table 9.1 The authors' compilation of primary categories of uses of assistive technologies to support ageing in place

SN	Specific use	Application to ageing in place[a]
Health and social care (Bente et al., 2023; Lariviere et al., 2021; Le Deist & Latouille, 2016)		
1	Homecare	People are supported with ATs to receive health and social care at home
2	Remote clinical care (by health professionals apart from physicians)	People are supported with ATs to remotely receive healthcare at home from clinical health professionals, enabling them to avoid long-term admission into care homes and facilities
3	Telemedicine and physician support	ATs enable people to engage with a doctor or medical professionals to remotely receive medical services, guidance, and monitoring
Mobility (Kononova et al., 2019; Thangavel et al., 2022; Tural et al., 2020)		
1	Move about at home	ATs support people with and without disabilities to move around at home, enabling them to exercise or perform ADLs
2	Move about in the community	People are supported by ATs to move beyond their homes, enabling them to perform IADLs (e.g., visiting the grocery shop).
3	Falls prevention	ATs enable people, especially frail older adults, to avoid falls when moving around and performing ADLs/IADLs
Sensory functioning (Lariviere et al., 2021; Sarfraz et al., 2017; Sprinzl & Riechelmann, 2010; Thangavel et al., 2022; Van Assche et al., 2023; Xie et al., 2020)		
1	Auditory functioning	ATs support individuals to hear while communicating or engaging with caregivers and others
2	Cognitive functioning and dementia care	ATs enable older adults to receive treatment of their neurodegenerative disorders (e.g. dementia). They also provide memory support
3	Vision aid	ATs (e.g. devices translating braille) enable older adults to see and perform ADLs/IADLs
4	Speech therapy and communication	ATs support people to speak clearly or understand verbal and nonverbal (sign) languages
Social functioning (Sarfraz et al., 2017; Sen et al., 2022; Thangavel et al., 2022; Xie et al., 2020; Zhao et al., 2022)		
1	In-person social engagement	ATs support older adults to participate in group activities (e.g. recreational and religious activities) with peers in-person
2	Remote/online social engagement	ATs support older adults to participate in online (Internet-based) chats and group activities (e.g. recreational and religious activities) with peers.
Well-being/preventive support (Popham et al., 2022; Sen et al., 2022; Wong et al., 2014)		
1	Exercise adherence	Devices enable older adults to track and maintain physical activity; some may remind older adults to exercise routinely or to engage with expert counsellors

(continued)

Table 9.1 (continued)

SN	Specific use	Application to ageing in place[a]
2	Medication adherence	ATs remind or guide individuals to stick to a preventive medication schedule prescribed by a physician
3	Health information and education seeking	Information technologies are used to acquire and seek health information online or from peers, experts, or public spaces
Safety and security (Badillo-Urquiola et al., 2020; Brims & Oliver, 2019; Daniel et al., 2009)[b]		
1	Safety at home	ATs (e.g. safety alarms) prompt older adults when they are exposed to health, social, or environmental risks at home, enabling them to ask for support
2	Safety in the community	ATs prompt people when they are exposed to health, social, or environmental risks in the community outside their homes, enabling them to take protective action
3	Safety online	ATs prompt people to identify, assess, or avoid social and health risks online

Note: *ATs* assistive technologies, *SN* serial number, *ADLs* activities of daily living, *IADLs* instrumental activities of daily living
[a]Ways in which assistive technologies may support people to maintain their engagement with life at home during the crises
[b]Particularly applies to ageing in place during the crises

life span. Though the DTA is at odds with the activity theory of ageing (ATA) and continuity theory of ageing (CTA), it is recognised as a valid framework explaining how people with a low socioeconomic status age (Asiamah, 2017).

The ATA and CTA posit that people can maintain resources and health-seeking behaviours by adapting previous life experiences (Asiamah et al., 2021; Havighurst, 1961). For example, a person can maintain the ability to exercise and utilise healthcare by making use of preserved skills and resources (e.g., income). This reasoning is congruent with the argument of the social cognitive theory (SCT) (White et al., 2012), which suggests that people can observe behaviours (e.g. exercise and healthcare utilisation) as they age and mimic those behaviours that are rewarding (Arkkukangas et al., 2021; White et al., 2012). The process of observing and performing rewarding behaviours is inherent in the adaptation of experiences around one's social ties including relatives, workmates, and caregivers. This idea overlaps with the Extended Unified Theory of Acceptance and Use of Technology (E-UTAUT) (Cimperman et al., 2016; Rahimi et al., 2018).

The E-UTAUT argues that the use of a technology partly depends on how much effort is needed and whether the use of the technology can improve health or any personal performance variable (Rahimi et al., 2018). The effort needed to use technologies depends on learning or adaptation of experiences with others. This thought recalls two other factors, namely, "social influence" and "doctor's opinion" within the E-UTAUT that can influence the use of technology (Cimperman et al., 2016);

both factors are implicit in people's social networks including their caregivers. Thus, both the E-UTAUT and SCT recognise the role of social ties in behaviours including the effective use of technologies. Moreover, SCT implies that the ability to use technologies can be influenced directly by others who provide guidance and skills and indirectly by those who perform behaviours as an example for others to emulate. This thought coupled with the import of the DTA suggests that the ability to use assistive technologies would fall over time, since income and social support tied to social networks can dwindle in ageing.

The socioemotional selectivity theory (SST) assumes that older adults are aware of how short their remaining life is and a need to use this time to pursue the most rewarding activities (Carstensen, 2021; Isaacowitz et al., 2003). This awareness is part of older adults' *future time perspective* (Carstensen, 2021), which necessitates the prioritisation of life choices in harmony with new goals accompanied by the ageing process. The negative aspect of the SST is that older adults limit their physical and social environment to their home and close relatives (e.g. grandchildren), respectively (Asiamah et al., 2021), which results in their social isolation and other risks (e.g. physical inactivity). This view of the SST suggests that older adults can lose social ties, income tied to economic activities, and functional ability needed to use assistive technologies in later life. The positive view of the SST is the possibility of older adults making choices (e.g. using technologies) that they perceive to be relevant to their longevity. Even older adults who are ambivalent about technologies may adopt and effectively use assistive technologies to maintain or improve their health (Chen, 2020), at least to have more time to savour their fondest relationships.

Deductively, ageing people face "demands" that may prevent them from using technologies across the life span; these demands involve a loss of functional and cognitive ability as well as social networks and income. Social isolation, physical inactivity, and poor health are consequences of having more demands. On the flip side, the ability to adapt past experiences, learn or mimic others, and retain resources that may decline with time are facilitators of assistive technology use. For people to use technologies effectively across the life span their resources should be more than their demands, which means that they must preserve relevant resources (e.g. income and social ties) and avoid demands over the life course. This idea draws on the job demands-resources (JD-R) theory (Demerouti et al., 2001a), which suggests that individual performance on a job depends on the ratio of demands to resources; if resources exceed demands, the individual's performance would be high, and vice versa. Demands are psychosocial and organisational factors (e.g. work pressure, emotional demands, and poor health) that counter individual performance (Demerouti et al., 2001b). Resources are factors such as training, job flexibility, and other motivational factors that facilitate high performance. This thinking forms the basis of our *technology demands-resources concept*.

9.4 Opportunities and Barriers to Technology Use Are Driven by a Demands-Resources Ratio

At the individual level, the main opportunity for using assistive technologies by ageing people is having more resources, which we refer to as "vitality". As the ATA and CTA suggest, the ability to adapt past experiences is tied to resources (e.g., social networks and income), so older people's ability to use assistive technologies would depend on vitality, which is the state of having enough resources that outplay demands. By this idea, we reason in harmony with the DTA that ageing people cannot avoid a chronological decline in resources, but those who can preserve enough resources through adaptative behaviour or human development programmes can use assistive technologies effectively and sustainably. Judging by the SCT and E-UTAUT, social networks including physicians can provide model behaviour or guidance by which people can use assistive technologies. The literature to date confirms that doctors and social ties (e.g. family members) facilitate sustained use of assistive technologies (Cimperman et al., 2016; Le Deist & Latouille, 2016).

Another opportunity for using assistive technologies is individuals' positive future time perspective mentioned earlier. Because older adults tend to highly value close relatives such as grandchildren in later life, their future time perspective impels them to do things that would allow them to spend more time with this group (Carstensen, 2021). As such, even those with ambivalence about technologies may use assistive technologies if this behaviour allows them to spend more time with closer relatives (Fokkema & Knipscheer, 2007; Pedrozo Campos Antunes et al., 2019). As our theoretical framework suggests, there are age-related challenges that should be avoided to maximise "vitality", especially during crises.

The opposite of vitality is "lethargy", which we use to refer to having more demands than resources so that the individual is less able and willing to use assistive technologies to maintain health. This individual condition is characterised by low income, physical and cognitive impairments, and the absence of supportive social ties (e.g. familial ties, carers, and friends) who provide model behaviours or guidance by which people sustainably use assistive technologies (Peek et al., 2014; Tural et al., 2020). Unlike the state of vitality, lethargy can be typical of groups in low socioeconomic areas where residents often live with economic deprivation and low access to services and social networks. Older adults in this state can find it more difficult to use assistive technologies, and their concerns and ambivalence about these technologies would be stronger. As reported in the literature (Cimperman et al., 2016; Fokkema & Knipscheer, 2007; Tural et al., 2020), their concerns may be about whether technologies are affordable, secure, trustworthy, and free of stigma.

Systematic reviews (Peek et al., 2014; Yusif et al., 2016) have reported barriers (e.g. high cost, low security, and low trustworthiness) to the use of assistive technologies among older adults, and these barriers are more likely to be faced by those in a state of lethargy. For instance, those who think they are unable to use information technologies due to their high cost are probably those without social support and gainful employment. Similarly, people who feel insecure about assistive technologies are likely those within cultures that are oblivious to technologies. Thus, barriers to the use of assistive technologies can be influenced by socioeconomic status, which is why eliminating socioeconomic inequalities is necessary.

Given the state of lethargy, it can be argued that the use of assistive technologies may not be sustained across the life span. Since the individual loses resources over time (Asiamah, 2017), people with fewer resources but more demands at a given time are less likely to accept and use assistive technologies across the life span. Thus, people in a lethargic state with possible socioeconomic disadvantages would realistically lose more resources in the ageing process and, therefore, lose the ability to use technologies. Worse yet, people's failure to sustainably use assistive technologies is more probable in contexts experiencing the crises. To illustrate, an electric power outage due to a tsunami can terminate the functionality of Internet-dependent assistive technologies. Flooding and storms can destroy robotic assistive technologies. For those in a lethargic state, fixing damaged technologies or getting them back to a usable form could be a mirage due to their socioeconomic disadvantages.

Some individuals may also live with equal proportions of resources and demands, a state we call *demands-resources balance*. This is a state of equilibrium where the positive impact of resources on the effective and sustained use of assistive technologies could be offset by demands. Consequently, people at this state, compared with those at the vitality state, are unlikely to maintain effective use of assistive technologies, especially during crises. Even so, human development programmes can modify or enhance resources over demands, enabling individuals to use assistive technologies to age in place. Corroborating this assertion are studies (Fotteler et al., 2022; Shapira et al., 2007; Van Assche et al., 2023) that have shown that human development interventions can enable people with demands (e.g. frailty, cognitive limitations, and low income) to adapt their resources (e.g. networks such as carers) to use assistive technologies.

9.5 Human Development Implications

Though some older adults, especially those in a state of vitality, can retain resources as they age, human development programmes (e.g. training, and mentoring) can be effective at maximising their ability to use assistive technologies (Fokkema & Knipscheer, 2007; Van Assche et al., 2023). Table 9.2 shows key human development programmes that have been used to encourage or facilitate the use of assistive technologies for ageing in place. Health promotion or education programmes can be used to positively influence population health behaviours including the use of assistive technologies to maintain health. Other programmes (e.g. controlled interventions) are costly and can, therefore, be applied only to a relatively small group but may be effective at testing the effectiveness of technologies and the health benefits of their use. Our theoretical framework also indicates that older adults may have unequal opportunities tied to resources to use assistive technologies. Research

Table 9.2 The authors' compilation of primary human development programmes and methods by which older adults can enhance their ability to use ATs

SN	Name (Reference[a])	Operational definition	Application to ageing in place[b]
1	Training (Kueider et al., 2012; Shapira et al., 2007)	In-service or on-the-job training aimed at improving older employees' skills and knowledge for using ATs	This is a way organisations can positively influence ageing in place by equipping their employees to use ATs to maintain health-seeking behaviours and engagement with life
2	Mentoring (Colón-Semenza et al., 2018; Giles et al., 2020)	A situation where a person learns to use ATs through the concerted guidance of an expert who could be a friend, workmate, relative or acquaintance	Mentoring is a way social ties can support ageing people to use technologies to perform ADLs/IADLs and to overcome or cope with problems caused by the crises
3	Formal education (Peek et al., 2014; Riddell & Song, 2017; Tural et al., 2020)	Learning to use ATs in the process of acquiring an academic qualification from an educational institution	Governments and educational institutions can infuse into curricula and pedagogies learning and training opportunities for effectively using ATs. Ideally, technology education should be compulsory for all and must be designed to support the use of health-supporting devices over the life course
4	Observation (Khasnabis et al., 2020; Tsai et al., 2015)	Learning to use ATs by watching and mimicking how others use them	Observation can take place at the workplace, in the community, or at home and can enable ageing people to acquire new skills and knowledge at no cost. This is a way people can improve competencies for using technologies to maintain health and engagement with life over the life course
5	Experimentation (Mariano et al., 2022; Tsai et al., 2015; Zhao et al., 2022)	Experimenting with or trying out new or unfamiliar technologies in one's curiosity or interest to know how they work	This is a suitable method in an emergency where a crisis has caused difficulties that people must overcome with new technologies. However, the risks involved are higher since the individual knows little about the dangers the technologies may pose
6	Controlled interventions (Fotteler et al., 2022; Laganá et al., 2011; Shapira et al., 2007; H. White et al., 2002)	Scientifically controlled or manipulated programmes (e.g. an experimental study) intended to help people to use technologies.	This can produce maximum impacts with minimal risks since they are carefully controlled and subjected to ethical procedures and safety scrutiny. It can be used to test technologies and their effectiveness for ageing in place

(continued)

Table 9.2 (continued)

SN	Name (Reference[a])	Operational definition	Application to ageing in place[b]
7	Public health education/promotion (Mariño et al., 2016; Stellefson et al., 2020; Zhang et al., 2023)	Efforts by governments and organisations (e.g. media campaigns and clinical counselling) to conscientise and guide the masses or specific groups to use technologies to maintain health and engagement with life at home	This can be effective at influencing larger populations at the county, national, or regional levels and can complement the other six programmes. For instance, people are more likely to experiment with technologies or participate in controlled interventions if they knew about the role of technologies in health through public health education or promotion

Note: *SN* serial number, *ATs* assistive technologies, *ADLs* activities of daily living, *IADLs* instrumental activities of daily living
[a]References showing examples of the method
[b]Ways to apply the method to support ageing in place

suggests (Riddell & Song, 2017; Shapira et al., 2007; Tural et al., 2020), nevertheless, that disparities in these opportunities can be eliminated with human development programmes aimed at improving skills and attitudes to technologies.

Sustained use of assistive technologies amidst crises requires a *technology-user fit*, which we define as the ability and willingness of people to use technologies to overcome a lethargic state or unique barriers to the maintenance of health over the life course. For people to overcome these barriers and maximise technology-user fit, assistive technologies must be user-friendly, interoperable (i.e. concurrently used by individuals and their caregivers), used with minimal cognitive and physical abilities, and must meet health and social needs. Since many older adults with physical and cognitive limitations depend on others to use assistive technologies (Cimperman et al., 2016; Hampel & Kunze, 2023), an interdisciplinary approach is necessary for developing these technologies for ageing in place, especially amidst crises.

Human development programmes should ideally be inclusive of all those involved in supporting older adults to use assistive technologies. This suggestion better applies to contexts experiencing crises where others (e.g. relatives and carers) may have to play a role to restore the functionality of assistive technologies disabled by the crises. Ageing in place only holds meaning if people age at home even if the neighbourhood is ravaged by the crises. Ideally, assistive technologies should be resilient against the crises; if possible, they should be independent of utilities (e.g. electric power and internet) whose supply can be cut by the crises. Users and their caregivers must be able to use assistive technologies to maintain access to services and relief providers in neighbourhoods experiencing the crises. Drawing on the above thoughts and our theoretical framework, we formulate the following postulates:

Postulate 1 – Individuals in a state of vitality, compared with those in a lethargic state, have more resources for using assistive technologies into later life, but human development programmes that enhance resources or the ability to use them can better sustain the effective use of assistive technologies to age at home in later life.

Postulate 2 – People with a lethargic state may have insufficient resources to maintain the use of assistive technologies, especially when there is a crisis, but human development programmes can enhance resources or the ability to use them, thereby enabling this group to maintain the effective use of assistive technologies to age at home into later life.

Postulate 3 – People with a balance between resources and demands (at the state of equilibrium) may have unstable resources and may, therefore, not use assistive technologies sustainably, but human development programmes can improve their resources over demands and technology-user fit, enabling them to use of assistive technologies to age at home into later life.

Postulate 4 – Human development programmes and other stakeholder efforts (e.g. government policies) can increase resources over demands by eliminating disparities in factors (e.g. education, income, employment, gender, and culture) influencing health and ageing. These efforts would enhance the state of vitality and move those in the states of equilibrium and lethargy into it.

Postulate 5 – Typical resources needed for the sustainable use of assistive technologies are in two categories. The primary category comprises social networks (e.g. relatives and carers) and income, whereas the second category encompasses personal conditions driven by the primary factors (i.e. physical and cognitive abilities) necessary for learning. Ageing people need both categories to maintain their use of assistive technologies, especially when crises occur. The more resources individuals have, the higher their ability to use assistive technologies effectively.

Postulate 6 – Individuals are better able to savour resources to effectively use assistive technologies amidst the crises at a higher technology-user fit, which ideally includes resilience over the crises. With resilience, older adults can use assistive technologies without excessively depending on others for support. The sustainable use of assistive technologies depends on whether users were involved in the design of these technologies.

As these postulates indicate, effective use of assistive technologies to age in place in contexts experiencing crises depends on collective effort, including human development programmes. Our thinking, nevertheless, has some limitations and implications for research.

9.6 Limitations and Future Research

Our postulates are only implications of our theoretical framework and may not be empirically supported. Their consideration in practice should ultimately be based on empirical research supporting them. This being so, future research testing the effect of technology-user fit on the sustainable use of assistive technologies to age in place is necessary. Researchers can assess whether this fit improves or preserves the ability to use assistive technologies over time in different groups of older adults. Future research examining the effects of different human development programmes on the ability to maintain the use of assistive technologies to engage with life at home in later life is needed. If possible, studies should analyse these effects between healthy older adults and those with cognitive and functional impairments. Similarly, longitudinal cluster randomised controlled trials can compare community-dwelling older adults experiencing the crises with those not experiencing any of the crises.

9.7 Conclusions

"Demands" experienced in ageing can be a barrier to the use of assistive technologies, but it is possible to preserve "resources" that facilitate sustained use of these technologies across the life span. Yet, the global crises can make it more difficult for individuals, especially those with physical and cognitive impairments, to preserve resources and maintain their use of assistive technologies to age in place. Even so, human development programmes can enable them to retain or increase their resources over their demands, attaining a state of vitality where they can sustain the use of assistive technologies to age at home despite the crises. A high technology-user fit is implicit in the vitality state where individuals can adapt optimum resources to overcome demands in using assistive technologies in later life, regardless of the crises. A technology-user fit overrides demands and the lethargic state through human development programmes. More so, human development may improve resources over demands for ageing people, including those at the state of equilibrium. Our framework suggests a need for people to be supported with human development programmes to use assistive technologies throughout the life span, rather than supporting them when they are already with disabilities and health problems.

Review Questions

1. In which ways can assistive technologies contribute to ageing in place? Discuss why assistive technologies may be more important during the crises.
2. How would you describe the technology-user fit, and why is it necessary for the sustainable use of assistive technologies?
3. Explain the three states about the individual's demands-resources ratio. Which of these states best supports the sustainable use of assistive technologies at home? Explain your choice.
4. Briefly explain six postulates connected to the technology demands-resources concept. How can each guide human development practice?
5. Enumerate six human development programmes or methods for enhancing the ability of individuals to use assistive technologies to age in place.

References

Arkkukangas, M., Eriksson, H. G., & Dension, E. (2021). Risk factors for fall-related injuries among community-dwelling men and women over 70 years of age, based on social cognitive theory: Results from a population study. *European Journal of Physiotherapy, 23*(4), 221–226. https://doi.org/10.1080/21679169.2019.1706631

Asiamah, N. (2017). Social engagement and physical activity: Commentary on why the activity and disengagement theories of ageing may both be valid. *Cogent Medicine, 4*(1), 1289664. https://doi.org/10.1080/2331205x.2017.1289664

Asiamah, N. (2022). *Walkable urban neighborhoods: The adverse effects of industrialization and climate change in developing countries* (pp. 247–262). Sustainable Urbanism in Developing Countries. https://doi.org/10.1201/9781003131922-16

Asiamah, N., Conduah, A. K., & Eduafo, R. (2021). Social network moderators of the association between Ghanaian older adults' neighbourhood walkability and social activity. *Health Promotion International*, 1–11. https://doi.org/10.1093/heapro/daaa156

Badillo-Urquiola, K., Chouhan, C., Chancellor, S., De Choudhary, M., & Wisniewski, P. (2020). Beyond parental control: Designing adolescent online safety apps using value sensitive design. *Journal of Adolescent Research, 35*(1), 147–175. https://doi.org/10.1177/0743558419884692

Barchielli, B., Cricenti, C., Gallè, F., Sabella, E. A., Liguori, F., Da Molin, G., Liguori, G., Orsi, G. B., Giannini, A. M., Ferracuti, S., & Napoli, C. (2022). Climate changes, natural resources depletion, COVID-19 pandemic, and Russian-Ukrainian war: What is the impact on habits change and mental health? *International Journal of Environmental Research and Public Health, 19*(19). https://doi.org/10.3390/ijerph191911929

Becker, S. E., Dickinson, J. I., Sullivan, K., & Cline, H. (2020). Baby boomer knowledge and stigma toward aging in place and residential assistive devices. *Journal of Interior Design, 45*(3), 43–61. https://doi.org/10.1111/joid.12176

Bente, B. E., Wentzel, J., Schepers, C., Breeman, L. D., Janssen, V. R., Pieterse, M. E., Evers, A. W. M., & Van Gemert-pijnen, L. (2023). Implementation and user evaluation of an eHealth technology platform supporting patients with cardiovascular disease in managing their health after a cardiac event: Mixed methods study. *JMIR Cardio, 7*, e43781. https://doi.org/10.2196/43781

Brims, L., & Oliver, K. (2019). Effectiveness of assistive technology in improving the safety of people with dementia: A systematic review and meta-analysis. *Aging and Mental Health, 23*(8), 942–951. https://doi.org/10.1080/13607863.2018.1455805

Brockie, L., & Miller, E. (2017). Older adults' disaster lifecycle experience of the 2011 and 2013 Queensland floods. *International Journal of Disaster Risk Reduction, 22*(August 2016), 211–218. https://doi.org/10.1016/j.ijdrr.2016.08.001

Carnemolla, P. (2018). Ageing in place and the internet of things – How smart home technologies, the built environment and caregiving intersect. *Visualization in Engineering, 6*(1). https://doi.org/10.1186/s40327-018-0066-5

Carstensen, L. L. (2021). Socioemotional selectivity theory: The role of perceived endings in human motivation. *Gerontologist, 61*(8), 1188–1196. https://doi.org/10.1093/geront/gnab116

Chen, K. (2020). Why do older people love and hate assistive technology? – An emotional experience perspective. *Ergonomics, 63*(12), 1463–1474. https://doi.org/10.1080/00140139.2020.1808714

Cimperman, M., Makovec Brenčič, M., & Trkman, P. (2016). Analyzing older users' home telehealth services acceptance behavior-applying an extended UTAUT model. *International Journal of Medical Informatics, 90*, 22–31. https://doi.org/10.1016/j.ijmedinf.2016.03.002

Claes, V., Devriendt, E., Tournoy, J., & Milisen, K. (2015). Attitudes and perceptions of adults of 60 years and older towards in-home monitoring of the activities of daily living with contactless sensors: An explorative study. *International Journal of Nursing Studies, 52*(1), 134–148. https://doi.org/10.1016/j.ijnurstu.2014.05.010

Colón-Semenza, C., Latham, N. K., Quintiliani, L. M., & Ellis, T. D. (2018). Peer coaching through mhealth targeting physical activity in people with parkinson disease: Feasibility study. *JMIR mHealth and uHealth, 6*(2). https://doi.org/10.2196/mhealth.8074

da Silva Francisco, A. A. (2017). 'Gerontogrowth' and population ageing in Africa and the Global AgeWatch Index. *Journal of the Economics of Ageing, 9*, 78–89. https://doi.org/10.1016/j.jeoa.2016.08.003

Daniel, K. M., Cason, C. L., & Ferrell, S. (2009). Emerging technologies to enhance the safety of older people in their homes. *Geriatric Nursing, 30*(6), 384–389. https://doi.org/10.1016/j.gerinurse.2009.08.010

Demerouti, E., Bakker, A. B., Nachreiner, F., & Schaufeli, W. B. (2001a). Demands resources model. *Journal of Applied Psychology, 86*(3), 499–512.

Demerouti, E., Nachreiner, F., Bakker, A. B., & Schaufeli, W. B. (2001b). The job demands-resources model of burnout. *Journal of Applied Psychology, 86*(3), 499–512. https://doi.org/10.1037/0021-9010.86.3.499

Dratsiou, I., Villacañas, O., Cooper, S., Isaris, P., Serras, M., Unzueta, L., & Fernandez-Carbajales, V. (2021). Assistive technologies for supporting wellbeing of older adults. *ACM International Conference Proceeding Series*, 338–341. https://doi.org/10.1145/3453892.3461352

Fokkema, T., & Knipscheer, K. (2007). Escape loneliness by going digital: A quantitative and qualitative evaluation of a Dutch experiment in using ECT to overcome loneliness among older adults. *Aging and Mental Health, 11*(5), 496–504. https://doi.org/10.1080/13607860701366129

Fotteler, M. L., Muhlbauer, V., Brefka, S., Mayer, S., Kohn, B., Holl, F., Swoboda, W., Gaugisch, P., Risch, B., Denkinger, M., & Dallmeier, D. (2022). The effectiveness of assistive technologies for older adults and the influence of frailty: Systematic literature review of randomized controlled trials. *JMIR Aging, 5*(2), 1–17. https://doi.org/10.2196/31916

Gamble, J. L., Hurley, B. J., Schultz, P. A., Jaglom, W. S., Krishnan, N., & Harris, M. (2013). Climate change and older Americans: State of the science. *Environmental Health Perspectives, 121*(1), 15–22. https://doi.org/10.1289/ehp.1205223

Giles, M., Baker, S. F., & Willis, J. M. (2020). Pre-service teachers' peer mentoring experience and its influence on technology proficiency. *Mentoring and Tutoring: Partnership in Learning, 28*(5), 602–624. https://doi.org/10.1080/13611267.2020.1859329

Hampel, K., & Kunze, F. (2023). *The older, the less digitally fluent? The role of age stereotypes and supervisor support*. 6–11.

Havighurst, R. J. (1961). No title. *The Gerontologist, 1*(1), 8–13. https://doi.org/10.1093/geront/1.1.8

Isaacowitz, D. M., Smith, T. B., & Carstensen, L. L. (2003). Socioemotional selectivity and mental health among trauma survivors in old age. *Ageing International, 28*(2), 181–199. https://doi.org/10.1007/s12126-003-1023-7

Khasnabis, C., Holloway, C., & MacLachlan, M. (2020). The digital and assistive technologies for ageing initiative: Learning from the GATE initiative. *The Lancet Healthy Longevity, 1*(3), e94–e95. https://doi.org/10.1016/S2666-7568(20)30049-0

Kononova, A., Li, L., Kamp, K., Bowen, M., Rikard, R. V., Cotten, S., & Peng, W. (2019). The use of wearable activity trackers among older adults: Focus group study of tracker perceptions, motivators, and barriers in the maintenance stage of behavior change. *JMIR mHealth and uHealth, 7*(4). https://doi.org/10.2196/mhealth.9832

Kueider, A. M., Parisi, J. M., Gross, A. L., & Rebok, G. W. (2012). Computerized cognitive training with older adults: A systematic review. *PLoS One, 7*(7). https://doi.org/10.1371/journal.pone.0040588

Laganá, L., Oliver, T., Ainsworth, A., & Edwards, M. (2011). Enhancing computer self-efficacy and attitudes in multi-ethnic older adults: A randomised controlled study. *Ageing and Society, 31*(6), 911–933. https://doi.org/10.1017/S0144686X10001340

Lariviere, M., Poland, F., Woolham, J., Newman, S., & Fox, C. (2021). Placing assistive technology and telecare in everyday practices of people with dementia and their caregivers: Findings from an embedded ethnography of a national dementia trial. *BMC Geriatrics, 21*(1), 1–13. https://doi.org/10.1186/s12877-020-01896-y

Le Deist, F., & Latouille, M. (2016). Acceptability conditions for telemonitoring gerontechnology in the elderly: Optimising the development and use of this new technology. *IRBM, 37*(5–6), 284–288. https://doi.org/10.1016/j.irbm.2015.12.002

Mariano, J., Marques, S., Ramos, M. R., Gerardo, F., da Cunha, C. L., Girenko, A., Alexandersson, J., Stree, B., Lamanna, M., Lorenzatto, M., Mikkelsen, L. P., Bundgård-Jørgensen, U., Rêgo, S., & de Vries, H. (2022). Too old for technology? Stereotype threat and technology use by older adults. *Behaviour and Information Technology, 41*(7), 1503–1514. https://doi.org/10.1080/0144929X.2021.1882577

Mariño, R. J., Marwaha, P., Barrow, S., & yan. (2016). Web-based oral health promotion program for older adults: Development and preliminary evaluation. *International Journal of Medical Informatics, 91*, e9–e15. https://doi.org/10.1016/j.ijmedinf.2016.04.002

McPhee, J. S., French, D. P., Jackson, D., Nazroo, J., Pendleton, N., & Degens, H. (2016). Physical activity in older age: Perspectives for healthy ageing and frailty. *Biogerontology, 17*(3), 567–580. https://doi.org/10.1007/s10522-016-9641-0

Menassa, M., Stronks, K., Khatmi, F., Roa Díaz, Z. M., Espinola, O. P., Gamba, M., Itodo, O. A., Buttia, C., Wehrli, F., Minder, B., Velarde, M. R., & Franco, O. H. (2023). Concepts and definitions of healthy ageing: A systematic review and synthesis of theoretical models. *EClinicalMedicine, 56*, 101821. https://doi.org/10.1016/j.eclinm.2022.101821

Methley, A. M., Campbell, S., Chew-Graham, C., McNally, R., & Cheraghi-Sohi, S. (2014). PICO, PICOS and SPIDER: A comparison study of specificity and sensitivity in three search tools for qualitative systematic reviews. *BMC Health Services Research, 14*(1). https://doi.org/10.1186/s12913-014-0579-0

Mois, G., & Beer, J. M. (2020). Robotics to support aging in place. In *Living with robots*. Elsevier. https://doi.org/10.1016/b978-0-12-815367-3.00003-7

Ntengwe, F. W. (2006). Pollutant loads and water quality in streams of heavily populated and industrialised towns. *Physics and Chemistry of the Earth, 31*(15–16), 832–839. https://doi.org/10.1016/j.pce.2006.08.025

Pani-Harreman, K. E., Bours, G. J. J. W., Zander, I., Kempen, G. I. J. M., & Van Duren, J. M. A. (2021). Definitions, key themes and aspects of "ageing in place": A scoping review. *Ageing and Society, 41*(9), 2026–2059. https://doi.org/10.1017/S0144686X20000094

Pedrozo Campos Antunes, T., Souza Bulle de Oliveira, A., Hudec, R., Brusque Crocetta, T., Ferreira de Lima Antão, J. Y., de Almeida Barbosa, R. T., Guarnieri, R., Massetti, T., Garner, D. M., & de Abreu, L. C. (2019). Assistive technology for communication of older adults: A systematic review. *Aging and Mental Health, 23*(4), 417–427. https://doi.org/10.1080/13607863.2018.1426718

Peek, S. T. M., Wouters, E. J. M., van Hoof, J., Luijkx, K. G., Boeije, H. R., & Vrijhoef, H. J. M. (2014). Factors influencing acceptance of technology for aging in place: A systematic review. *International Journal of Medical Informatics, 83*(4), 235–248. https://doi.org/10.1016/j.ijmedinf.2014.01.004

Popham, S., Burq, M., Rainaldi, E. E., Shin, S., Dunn, J., & Kapur, R. (2022). Validation study for an algorithm to classify real-world ambulatory status from a wearable device using multimodal and demographically diverse data (preprint). *JMIR Biomedical Engineering, 8*, 1–14. https://doi.org/10.2196/43726

Pruchno, R. (2019). Technology and aging: An evolving partnership. In *Gerontologist* (Vol. 59, Issue 1, pp. 1–5). Gerontological Society of America. https://doi.org/10.1093/geront/gny153

Rahimi, B., Nadri, H., Afshar, H. L., & Timpka, T. (2018). A systematic review of the technology acceptance model in health informatics. *Applied Clinical Informatics, 9*(3), 604–634. https://doi.org/10.1055/s-0038-1668091

Riddell, W. C., & Song, X. (2017). The role of education in technology use and adoption: Evidence from the Canadian workplace and employee survey. *ILR Review, 70*(5), 1219–1253. https://doi.org/10.1177/0019793916687719

Sarfraz, M. S., Constantinescu, A., Zuzej, M., & Stiefelhagen, R. (2017). A multimodal assistive system for helping visually impaired in social interactions. *Informatik-Spektrum, 40*(6), 540–545. https://doi.org/10.1007/s00287-017-1077-7

Sen, K., Prybutok, G., & Prybutok, V. (2022). The use of digital technology for social wellbeing reduces social isolation in older adults: A systematic review. *SSM – Population Health, 17*(December 2021), 101020. https://doi.org/10.1016/j.ssmph.2021.101020

Sezgin, D., O'Donovan, M., Cornally, N., Liew, A., & O'Caoimh, R. (2019). Defining frailty for healthcare practice and research: A qualitative systematic review with thematic analysis. *International Journal of Nursing Studies, 92*, 16–26. https://doi.org/10.1016/j.ijnurstu.2018.12.014

Shapira, N., Barak, A., & Gal, I. (2007). Promoting older adults' well-being through Internet training and use. *Aging and Mental Health, 11*(5), 477–484. https://doi.org/10.1080/13607860601086546

Sprinzl, G. M., & Riechelmann, H. (2010). Current trends in treating hearing loss in elderly people: A review of the technology and treatment options – A mini-review. *Gerontology, 56*(3), 351–358. https://doi.org/10.1159/000275062

Stellefson, M., Paige, S. R., Chaney, B. H., & Chaney, J. D. (2020). Evolving role of social media in health promotion: Updated responsibilities for health education specialists. *International Journal of Environmental Research and Public Health, 17*(4). https://doi.org/10.3390/ijerph17041153

Takatori, K., Matsumoto, D., Miyazaki, M., Yamasaki, N., & Moon, J.-S. (2018). Relationship between self-perceived age and social activity in older Japanese adults: The KAGUYA Study. *Health, 10*(11), 1459–1473. https://doi.org/10.4236/health.2018.1011112

Thangavel, G., Memedi, M., & Hedström, K. (2022). Customized information and communication technology for reducing social isolation and loneliness among older adults: Scoping review. *JMIR Mental Health, 9*(3), 1–14. https://doi.org/10.2196/34221

Tsai, H.-y. S., Shillair, R., Cotten, S. R., Winstead, V., & Yost, E. (2015). Getting grandma online: Are tablets the answer for increasing digital inclusion for older adults in the U.S.? *Educational Gerontology, 41*(10), 695–709. https://doi.org/10.1080/03601277.2015.1048165

Tural, E., Lu, D., & Cole, D. A. (2020). Factors predicting older adults' attitudes toward and intentions to use stair mobility assistive designs at home. *Preventive Medicine Reports, 18*(August 2019), 101082. https://doi.org/10.1016/j.pmedr.2020.101082

Van Assche, M., Moreels, T., Petrovic, M., Cambier, D., Calders, P., & Van de Velde, D. (2023). The role of a socially assistive robot in enabling older adults with mild cognitive impairment to cope with the measures of the COVID-19 lockdown: A qualitative study. *Scandinavian Journal of Occupational Therapy, 30*(1), 42–52. https://doi.org/10.1080/11038128.2021.2009560

White, H., McConnell, E., Clipp, E., Branch, L. G., Sloane, R., Pieper, C., & Box, T. L. (2002). A randomized controlled trial of the psychosocial impact of providing internet training and access to older adults. *Aging and Mental Health, 6*(3), 213–221. https://doi.org/10.1080/13607860220142422

White, S. M., Wójcicki, T. R., & McAuley, E. (2012). Social cognitive influences on physical activity behavior in middle-aged and older adults. *Journals of Gerontology – Series B Psychological Sciences and Social Sciences, 67B*(1), 18–26. https://doi.org/10.1093/geronb/gbr064

Wong, C. K. M., Yeung, D. Y., Ho, H. C. Y., Tse, K. P., & Lam, C. Y. (2014). Chinese older adults' internet use for health information. *Journal of Applied Gerontology, 33*(3), 316–335. https://doi.org/10.1177/0733464812463430

Xie, B., Charness, N., Fingerman, K., Kaye, J., Kim, M. T., & Khurshid, A. (2020). When going digital becomes a necessity: Ensuring older adults' needs for information, services, and social inclusion during COVID-19. *Journal of Aging and Social Policy, 32*(4–5), 460–470. https://doi.org/10.1080/08959420.2020.1771237

Yusif, S., Soar, J., & Hafeez-Baig, A. (2016). Older people, assistive technologies, and the barriers to adoption: A systematic review. *International Journal of Medical Informatics, 94*, 112–116. https://doi.org/10.1016/j.ijmedinf.2016.07.004

Zhang, B., Ying, L., Khan, M. A., Ali, M., Barykin, S., & Jahanzeb, A. (2023). Sustainable digital marketing: Factors of adoption of M-technologies by older adults in the Chinese market. *Sustainability, 15*(3), 1972. https://doi.org/10.3390/su15031972

Zhao, Y., Zhang, T., Dasgupta, R. K., & Xia, R. (2022). Narrowing the age-based digital divide: Developing digital capability through social activities. *Information Systems Journal, June 2022*, 268–298. https://doi.org/10.1111/isj.12400

Chapter 10
Infectious Diseases and Healthy Ageing: Making the Case for a 15-Minute City

Evelyn N. Alvarez

Acronyms

AD	Alzheimer's disease
CBO	Community-based organization
COVID-19	Coronavirus disease
CSV	Comma-separated values
MeSH	Medical Subject Headings
VR	Virtual reality

10.1 Introduction

When COVID-19 emerged as a global public health threat, we witnessed the impacts it had on mental health and social interaction. Studies demonstrated that the pandemic was linked to reduced quantity and quality of social interactions (Finlay et al., 2022; Kotwal et al., 2021) with deeper disproportionate impacts on older adults (Kotwal et al., 2021). Significant public health gains have been made with the findings that physical activity can be protective of cognitive decline in older adults (Hörder et al., 2018; Rovio et al., 2005) leading to recommendations and interventions that focus on increasing outdoor physical activity in older adults (Roe et al., 2020). With the closure of public parks and beaches and the encouragement from public health officials to *physically distance*, the importance of physical activity as a health promoting activity faded away. Hence, there arose what we call a *public*

E. N. Alvarez (✉)
Department of Public Health, California State University, Los Angeles,
Los Angeles, CA, USA
e-mail: evelyna@calstatela.edu

N. Asiamah et al. (eds.), *Sustainable Neighbourhoods for Ageing in Place*,
https://doi.org/10.1007/978-3-031-41594-4_10

health dilemma during a pandemic emergency – on the one hand, *physically distancing* was an evidence-based way to avoid getting sick and spreading a novel coronavirus (Talic et al., 2021), yet this meant that many people would not be able to exercise in public to gain or maintain health benefits.

Assumptions were made that all people, regardless of age and access to resources, would be able to transform the way they learned, worked, and engaged in necessary physical activity by retreating to the indoors and relying on more remote, technological ways of interacting with the outside world. For many older adults who relied on the outdoors for physical activity (jogs, walks, strength training), being able to adapt their routine indoors was dependent on whether they would have indoor environments conducive to physical activity. Furthermore, as gyms and gym classes closed due to the pandemic, many older adults who used that as a way to engage in social physical activity were left with the prospect of navigating online videos for physical activity (Greenwood-Hickman et al., 2021). Some health insurance carriers, such as Kaiser Permanente, offer physical exercise and health programming that was made available in an online format during the pandemic onset, and some older adults were able to put in the time to learn how to use Zoom (Greenwood-Hickman et al., 2021). However, some older adults reported feeling a loss of much needed social interaction as their group exercise classes closed, and their work and volunteer work, religious services, and community-facing errand routines were impacted (Greenwood-Hickman et al., 2021).

In many ways, older adults were not just the canary in the coal mine for the bandwidth of our health systems but also for the resilience of our ageing in place public health identity. A more nuanced *public health dilemma* that specifically applies to older adults therefore emerges –physical activity, social engagement, and exposure to green space are particularly important for older adults in protecting against cognitive decline, yet emerging infectious diseases that become full-blown pandemics threaten the very idea of ageing in place. A richness of resources is available to address this dilemma; however, public health messaging, city planning, and access to these resources need to be critically addressed.

10.2 Review Methodology and Key Findings

This section provides a summary of literature search methodology and findings from the search regarding how emerging infectious diseases, such as COVID-19, impact ageing in place.

10.2.1 Search Methodology

Our search aimed to identify published (completed) journal research articles on how the COVID-19 pandemic impacted ageing in place for older adults. The databases we searched were Scopus and PubMed. We performed the search in the second week of April of 2023. The search was based on primary key terms (i.e. ageing, ageing in place, COVID-19, resources, health, environment, older adults, infectious disease) searched using the Boolean operators 'OR' to include synonymous terms and similar variations of terms and 'AND' to combine all search terms. Synonyms of the key terms were identified using MeSH (Medical Subject Headings) terms on PubMed. Brackets '[]' and '()' were used in concert with Boolean operators to qualify search rules. Lastly, when available, filters were used to limit searches to articles that have been published (completed), are in English, and included humans as subjects (non-animal lab experiments). In addition, we limited our search to research articles and reviews. A total of 523 articles were downloaded. After applying the inclusion criteria, 103 documents were identified as relevant. The comma-separated values (CSV) files from both Scopus and PubMed were uploaded to Bibliometrix's web interface, Biblioshiny, which we used as a bibliometric tool (accessed through R) to analyse all 103 articles (Aria & Cuccurullo, 2017).

10.2.2 Infectious Diseases and Ageing in Place

There is a dearth of tools that assess quality of life of older adults ageing in place (Vanleerberghe et al., 2017). Specifically mentioned throughout the literature, cost-benefit analyses of closing beaches, parks, and other outdoor recreational areas during pandemics need to be conducted to assess the pros and cons of policies. Spending less time outdoors is associated with higher depression and anxiety scores (Young et al., 2022). Older adults who live in cities with more walkability infrastructure experience positive mental health and social health benefits. One scale has been proposed to measure walkability as a product of the built environment (Asiamah et al., 2020).

During the peak points of the COVID-19 pandemic, older adults were in the highest mortality group; however, they were not the principal spreaders of the virus (Golubev & Sidorenko, 2020). The idea of social distancing in older adults, especially in light of outdoor spaces, has been viewed as a 'discriminatory measure' (Golubev & Sidorenko, 2020; Reynolds, 2020; Jen et al., 2021). Table 10.1 shows eight categories that are representative of the eight most dominant facets of COVID-19's impact on ageing in place as identified in the literature. 'Meaningful participation in the community' encapsulates COVID-19's impact on older adults' ability to continue their volunteer work, whether at community organizations, schools, or churches, which brings value to their lives. Socializing and friendships are particularly important especially in the context of cognitive decline and prevention of Alzheimer's disease (AD). Mental health and loneliness were the most commonly investigated impacts of COVID-19 on healthy ageing, one that was highly correlated with lack of access to the outdoors (through stay-at-home orders). Exercise and mobility were also heavily examined in the literature especially in connection with policies that prevented older adults from exercising in the outdoors. Walkability in cities was examined and determined to be closely linked to physical activity and mental health.

Table 10.1 The authors' compilation of primary facets of COVID-19's impact on ageing in place as found in the literature

Domain	Relevant literature
Meaningful participation in the community	Pestine-Stevens and Greenfield (2020), DeLange Martinez et al. (2020), Finlay et al. (2022), Principi et al. (2022), Tan and Idris (2023)
Socializing/ friendships	Emerson (2020), Richards et al. (2021), Tiilikainen et al. (2021), Chen et al. (2022a), Asiamah et al. (2021)
Mental health and loneliness	Hsu et al. (2020), Mikal et al. (2021), Krendl and Perry (2021), Herron et al. (2021), O'Shea et al. (2021), Lennon et al. (2022), Masters et al. (2022), Joseph et al. (2022), Herron et al. (2022), Pinazo-Hernandis et al. (2022), Carpagnano et al. (2021), Dejvajara et al. (2022)
Agency	Piccoli et al. (2020)
Accessing healthcare	McDonald et al. (2021)
Exercise and mobility	Rantanen et al. (2021), Angelo et al. (2022), Aung et al. (2022a), Liu and Lu (2022), Zingmark et al. (2022), Asiamah et al. (2021, 2022)
Technology usage	Albers et al. (2022), Ghosh (2022), Perdana and Mokhtar (2022), Szczuka et al. (2021)
Self-neglect	Manthorpe et al. (2022)

Technology usage was an important modulator of COVID-19's imposed limitations and an adaptive tool in connecting with family and friends, provided older adult's existing proficiency or willingness to learn. As the COVID-19 pandemic forced populations to navigate web-based social interactions, some without training, this inevitably created stress, particularly in older adults who predominantly are stated in the literature to have initially struggled with remote applications. Nevertheless, resiliency was witnessed across many studies, indicating the willingness and dedication of older adults to adapt to new situations that require using technology. In many cases, educational interventions on online technology usage (Zoom, telehealth, etc.) targeted at older adults showed promise in helping older adults to stay connected during pandemics. Accessing healthcare (face-to-face), developing agency over one's own health decisions, and self-neglect were other important facets of healthy ageing that were examined in the literature.

We distinguish between primary facets of COVID-19's impacts on ageing in place (Table 10.1) *and* primary categories of possible interventions to promote ageing in place during emerging pandemic threats (Table 10.2). Our findings indicate some lack of synchrony between facets of COVID-19's impacts and categories of possible interventions to address healthy ageing during pandemics. The most prevalent category in Table 10.1 is 'Mental Health and Loneliness', yet this was not the most commonly addressed category of interventions depicted in Table 10.2. The most commonly addressed categories of interventions relating to COVID-19's impact on older adults were city planning and 'involvement of social workers and/ or connecting clinical and community-based organizations (CBOs)'. In particular, there was a focus on the importance of designing cities in ways that are aligned with healthy ageing to ensure older adults have unfettered access to resources.

Table 10.2 The authors' compilation of primary categories of possible interventions to promote ageing in place during COVID-19 as found in the literature

Intervention	Relevant literature
Involvement of social workers and/or connecting clinical and CBOs[a]	Pestine-Stevens and Greenfield (2020), Smith et al. (2020), Hoang et al. (2021), Park and Go (2021), Meuser et al. (2022), Dabelko-Schoeny et al. (2022), Yeh et al. (2022), Naruse et al. (2023), Chen et al. (2022b), Fang et al. (2022)
Access to wearables, apps, VR[a], and smart systems	Muheidat and Tawalbeh (2020), Shao and Lee (2020), Van Hees et al. (2021), Gupta et al. (2021), Stradford et al. (2021), Nagajayanthi et al. (2021), Pech et al. (2022), Soubutts et al. (2022)
Tele-therapy and medicine	Kordova and Keisari (2020), Sekhon et al. (2021), Fournier et al. (2021a, b), Okoye et al. (2021), Boone and Shammash (2022), Hill et al. (2021)
Medication adherence	Esposti et al. (2020), Zhu et al. (2022), Washington et al. (2023)
Creating an emergency preparedness plan	Shang et al. (2020), Bell et al. (2022), Greer et al. (2023)
Older adult minority health	Angel and Mudrazija (2020), Garcia et al. (2021), Spellman and Ntiri (2022), Tai-Seale et al. (2023)
Low-income older adults	Buffel et al. (2020), Calo (2022), Greer et al. (2023)
LGBTQ[a] older adults	Prasad et al. (2022)
Creating a feeling of autonomy	Nurain et al. (2021)
City planning	Brandén et al. (2020), Marston et al. (2020), Phillips et al. (2021), Rosa-Jiménez and Nebot (2021), Gaglione et al. (2021, 2022), Chau et al. (2022), Figurska et al. (2022), Fang et al. (2022), James (2023)
Mass transit	Frantál et al. (2020)
Social robotics	Miller and McDaniel (2021), Perotti and Strutz (2022), Sut et al. (2022)
Experiencing nature	Lõhmus et al. (2021)
Human-centred approach	Resendiz-Villasenor et al. (2021), Funk et al. (2023), Washington et al. (2023)
Religion	Amanat et al. (2021)
Exercise	Aung et al. (2022b), Di Castelnuovo et al. (2021), Dürst et al. (2022), Kohn et al. (2023)

[a]*CBO* community-based organization, *VR* virtual reality, *LGBTQ* lesbian, gay, bisexual, transgender, queer

As indicated in Table 10.2, access to wearables, apps, virtual reality (VR), and smart systems as well as tele-therapy and medicine played significant roles in the interventions found in the literature. Most involved an educational component that would teach subsets of older adults how to use them, including some as simple as learning to type an email to a relative to something more advanced such as a sensor put under a carpet to measure gait and provide information on fall risks. Many older adults reported feeling frustrated at the initial stages of the learning process but demonstrated resiliency in continuing with the learning process. Most studies demonstrated that many older adults are able to learn how to use these types of technologies with time, patience, and instructional resources.

10.3 Ageing in Place and the 15-Minute City

In 2018, the American Geriatrics Society published a White Paper on Healthy Ageing (Batsis et al., 2021) that proposed the following domains as being core to healthy ageing of adults:

1. Health promotion
2. Injury prevention
3. Cognitive health
4. Physical health
5. Mental health
6. Social health

COVID-19 has challenged all of the aforementioned domains. It is therefore crucial to strengthen these six areas to make them more pandemic-proof. Emerging infectious diseases are not the only factor that interrupts healthy ageing. Climate change and infectious diseases are synergistically a growing threat to livelihoods. The concept of a '15-minute city' (Allam et al., 2022; Khavarian-Garmsir et al., 2023) could theoretically address all of these domains and could potentially stand the test of an emerging infectious disease. We must have access to more sustainable ways of living, and therefore, we believe that sustainability is synonymous with healthy ageing. The "15-minute city" concept has seven underlying principles according to Khavarian-Garmsir et al., 2023:

1. Proximity
2. Density
3. Diversity
4. Digitalization
5. Human scale urban design
6. Connectivity
7. Flexibility

Proximity as a principle refers to having access to key elements of a healthy city (healthcare, recreation, markets, etc.) within a 15-minute walk or bicycle ride. Density pertains to population density that could provide the basis for mass transit demand and job creation, which would decrease dependency on private cars and promote sustainability. Diversity alludes to increasing neighbourhood cultural, ethnic, economic, and social diversity. Digitalization is a key aspect of the '15-minute city' that leverages smart city technologies to help connect community members with each other. Human scale urban design calls for the development of existing urban areas to reduce pressure on suburban areas, which would be in line with promoting the usage of parks. Connectivity is a key aspect of the '15-minute city' that targets isolation by proposing infrastructure that enables walking, cycling, and public transit efficiency. Flexibility is a dimension that advocates for the multi-purpose use of various public spaces (e.g., creating vaccination sites on college campuses, converting parking spaces into outdoor seating, etc.), something we learned to do as a result of the COVID-19 pandemic.

The concept of the 15-minute city, the idea that living in a city where everything one needs (markets, restaurants, schools, hospitals, clinics, libraries, parks, etc.) is accessible via foot, bike, or public transit within 15 minutes of one's home, should be entrenched in discussions of healthy ageing, especially in the context of infectious diseases. 15-minute cities concentrate resources in small urban pockets, which are important during pandemics that impact access to resources, in particular by older adults. Largely absent from the literature search we conducted, the concept of 15-minute cities could be a promising 'ageing in place' research area that could help city planners and local governments design better places to live and age in place.

We advocate for local governments to focus on agenda setting that would prioritize the transformation of neighbourhoods into 15-minute cities. Whether it's a 5-, 10-, 15-, or 20-year plan, local governments should be motivated to head toward a more sustainable (with fewer greenhouse gas emissions), ageing-friendly, and pandemic resilient type of city design. We believe the co-benefits of 15-minute cities are multi-fold and address the six aforementioned core domains of healthy ageing in the following ways:

1. Being able to walk or ride a bike to run errands (doctor's appointments, grocery shopping, socializing, etc.) promotes overall health.
2. Being more mobile and active strengthens the body to prevent injuries.
3. Being able to more conveniently attend community engagement events promotes cognitive health.
4. Burning calories and gaining muscle as one runs errands promote physical health.
5. Being in the outdoors and getting around by foot or by bike allow for opportunities of peaceful self-reflection that may promote mental health.
6. Proximity to restaurants, cafes, and other meeting places allows for one to more conveniently spend quality time with family and friends and even create new friends in busy public spaces, which promotes social health.

The concept of a 15-minute city is a concept that not only captures the essence of a sustainable city but simultaneously also solves the healthy ageing problem.

10.4 An Ageing in Place Ecosystem That Is Pandemic-Proof

Creating an ageing in place ecosystem that is resilient in times of infectious disease outbreaks and other emergencies is critical. As digitalization makes for new innovative products, we should address older adults' access to these resources within the context of how it can help ageing in place. Therefore, the term 'ageing in place ecosystem' would incorporate technologies along with educational interventions to promote their usage in older adults. This term also reflects the highly multidisciplinary nature of the healthy ageing research field.

It is of crucial importance that during a pandemic, efforts are made to keep the outdoors (parks, beaches, recreational areas) open to maintain exercise routines but also to protect associated mental health. Therefore, during pandemics like COVID-19, we postulate the following:

1. Quality of life needs to be considered in a risk-benefit analysis when it comes to the closure of public outdoor spaces like parks, beaches, and recreational areas.

Since our findings indicate some lack of synchrony between facets of COVID-19's impacts and categories of possible interventions to address healthy ageing during pandemics, we postulate the following:

2. More research needs to be conducted on designing interventions that address mental health and loneliness in older adults during pandemics to ensure healthy ageing in place is not disrupted.

By inference, since most commonly addressed categories of interventions relating to COVID-19's impact on older adults were city planning and 'involvement of social workers and/or connecting clinical and community-based organizations', we postulate the following:

3. Interventions that target the aforementioned categories (i.e., city planning and involvement of social workers, etc.) need be actively translated into action via policies and initiatives.

The built environment and sustainability are inextricably intertwined with healthy ageing. Table 10.3 highlights some of our suggestions for ecosystem characteristics

Table 10.3 The authors' suggestions on what constitutes a healthy ageing in place ecosystem that is resilient to emerging pandemic threats and their links to '15-minute city' principles

Ageing in place ecosystem attributes	'15-minute city' dimensions
1. Access to a stable support network (family, friends, volunteers, etc.)	Digitalization, connectivity
2. Access to local restaurants/coffee shops with outdoor seating	Human scale urban design, flexibility
3. Access to a local library	Digitalization
4. Access to safe outdoor areas for walking/exercising	Density, flexibility
5. Access to technology (and training)	Digitalization
6. Access to parks	Density, proximity
7. Access to the beach (if nearby)	Human scale urban design
8. Access to local mass transit	Digitalization, connectivity
9. Access to religious services (if applicable)	Connectivity, diversity
10. Access to community events	Connectivity, diversity
11. Access to volunteering opportunities	Connectivity, diversity
12. Access to green spaces	Density, proximity
13. Access to healthy markets	Connectivity, proximity
14. Access to clinics and other medical services	Density, proximity
15. Access to a hospital	Density, proximity
16. Access to a gym with an outdoor space	Density, proximity

of healthy ageing and their connections to '15-minute city' principles. If most of the intervention research related to ageing in place during pandemics is pointing toward city planning and connecting social workers, clinics, and CBOs, then creating the infrastructure to make this happen should be analogous to developing a 'Millennium Development Goal'-type culture. It should permeate policy discussions at the federal and local levels so that it can become part of the national agenda and eventually part of a country's *national identity*. Therefore, creating sustainable cities results in co-benefits that translate to healthy ageing. A city that ages well is also a city that is well-prepared for emerging pandemic threats.

10.5 Limitations and Future Research

Our postulates are only recommendations based on our findings and may not be empirically supported. We further suggest that more empirical research be conducted to be able to qualify and quantify these postulates. We identified the potential of a key concept such as the 15-minute city to help us move toward agenda setting that prioritizes ageing in place in city planning discussions. We propose referring to healthy ageing in cities as 'ageing in place ecosystems' to further exemplify the interconnectedness of multidisciplinary factors that impact older adults. Future research should examine the viability of the 15-minute city as a way to positively address healthy ageing and to make cities more pandemic-proof. Additionally, more studies need to determine what terms or messaging in the 'healthy ageing' research field has more traction or public appeal to help make ageing discussions a part of our everyday culture.

10.6 Conclusions

Barriers to healthy ageing can be found in the built environment itself. Emerging infectious diseases such as COVID-19 can threaten lives and livelihoods, especially livelihoods that exhibit healthy ageing practices. Ensuring that factors such as physical activity, social engagement, and mental health that positively impact ageing are preserved or even enhanced as much as possible during pandemics should be part of every country's national agenda. Our research reveals a disconnect between the most investigated healthy ageing facets impacted by COVID-19 and the most common category addressed by interventions geared toward healthy ageing. We propose for research intervention efforts to more closely align with identified areas of need. In light of frameworks that produce multiple public health co-benefits, such as the 15-minute city, it is appropriate to examine their potential and adaptability in addressing healthy ageing. Lastly, determining what terms and messaging gain more public acceptability can help entrench healthy ageing into the culture of everyday life.

Review Questions

1. In what ways do infectious diseases, such as COVID-19, impact healthy ageing in place?
2. Describe the two most dominant research facets of COVID-19's impact on ageing in place.
3. Describe the two most dominant categories of possible interventions that promote ageing in place during COVID-19.
4. Discuss how there is a lack of synchrony between the most commonly investigated impact of COVID-19 as it pertains to healthy ageing and the most common category addressed by interventions geared toward healthy ageing.
5. How does the concept of a '15-minute city' relate to healthy ageing?
6. Briefly explain three postulates connected to ageing in place during emerging infectious diseases.
7. Name three characteristics of an ageing in place ecosystem that are pandemic-proof.

References

Albers, E. A., Mikal, J., Millenbah, A., Finlay, J., Jutkowitz, E., Mitchell, L., et al. (2022). The use of technology among persons with memory concerns and their caregivers in the United States during the COVID-19 pandemic: Qualitative study [Article]. *JMIR Aging, 5*(1), Article e31552. https://doi.org/10.2196/31552

Allam, Z., Nieuwenhuijsen, M., Chabaud, D., & Moreno, C. (2022). The 15-minute city offers a new framework for sustainability, liveability, and health. *Lancet Planet Health, 6*(3), e181–e183. https://doi.org/10.1016/S2542-5196(22)00014-6

Amanat, M., Rezaei, N., Roozbeh, M., Shojaei, M., Tafakhori, A., Zoghi, A., et al. (2021). Neurological manifestations as the predictors of severity and mortality in hospitalized individuals with COVID-19: A multicenter prospective clinical study. *BMC Neurology, 21*(1), 116. https://doi.org/10.1186/s12883-021-02152-5

Angel, J. L., & Mudrazija, S. (2020). Local government efforts to mitigate the novel coronavirus pandemic among older adults [Article]. *Journal of Aging and Social Policy, 32*(4–5), 439–449. https://doi.org/10.1080/08959420.2020.1771240

Angelo, F. D. D. A., Fonseca, F. D. S., Farah, B. Q., de Araújo, R. C., Cavalcante, B. R., Beltrão, N. B., & Torres Pirauá, A. L. (2022). Changes in physical functioning and fall-related factors in older adults due to COVID-19 social isolation [Article]. *Canadian Geriatrics Journal, 25*(3), 240–247. https://doi.org/10.5770/cgj.25.591

Aria, M., & Cuccurullo, C. (2017). bibliometrix: An R-tool for comprehensive science mapping analysis. *Journal of Informetrics, 11*(4), 959–975. https://doi.org/10.1016/j.joi.2017.08.007

Asiamah, N., Kouveliotis, K., Eduafo, R., & Borkey, R. (2020). Assessing the maturity of community-level built environments for older adults' active behaviors: Highlights from a psychometric test. *Advances in Gerontology, 33*(1), 153–158.

Asiamah, N., Conduah, A. K., & Eduafo, R. (2021). Social network moderators of the association between Ghanaian older adults' neighbourhood walkability and social activity. *Health Promotion International, 36*(5), 1357–1367. https://doi.org/10.1093/heapro/daaa156

Asiamah, N., Vieira, E. R., Kouveliotis, K., Gasana, J., Awuviry-Newton, K., & Eduafo, R. (2022). Associations between older African academics' physical activity, walkability and mental health: A social distancing perspective. *Health Promotion International, 37*(2). https://doi.org/10.1093/heapro/daab093

Aung, M. N., Koyanagi, Y., Ueno, S., Tiraphat, S., & Yuasa, M. (2022a). Age-friendly environment and community-based social innovation in Japan: A mixed-method study [Article]. *Gerontologist, 62*(1), 89–99. https://doi.org/10.1093/geront/gnab121

Aung, M. N., Moolphate, S., Aung, T. N. N., Koyanagi, Y., Kurusattra, A., Chantaraksa, S., et al. (2022b). Effectiveness of a community-integrated intermediary care (CIIC) service model to enhance family-based long-term care for Thai older adults in Chiang Mai, Thailand: A cluster-randomized controlled trial TCTR20190412004. *Health Res Policy Syst, 20*(Suppl 1), 110. https://doi.org/10.1186/s12961-022-00911-5

Batsis, J. A., Daniel, K., Eckstrom, E., Goldlist, K., Kusz, H., Lane, D., et al. (2021). Promoting healthy aging during COVID-19. *Journal of the American Geriatrics Society, 69*(3), 572–580. https://doi.org/10.1111/jgs.17035

Bell, S. A., Krienke, L., Brown, A., Inloes, J., Rettell, Z., & Wyte-Lake, T. (2022). Barriers and facilitators to providing home-based care in a pandemic: Policy and practice implications [Article]. *BMC Geriatrics, 22*(1), Article 234. https://doi.org/10.1186/s12877-022-02907-w

Boone, E. J., & Shammash, J. B. (2022). Virtual hospitalization at home after the Covid-19 pandemic: Time to consider our future community healthcare options [Note]. *Journal of Primary Care and Community Health, 13*. https://doi.org/10.1177/21501319221112586

Brandén, M., Aradhya, S., Kolk, M., Härkönen, J., Drefahl, S., Malmberg, B., et al. (2020). Residential context and COVID-19 mortality among adults aged 70 years and older in Stockholm: A population-based, observational study using individual-level data [Article]. *The Lancet Healthy Longevity, 1*(2), e80–e88. https://doi.org/10.1016/S2666-7568(20)30016-7

Buffel, T., Doran, P., Goff, M., Lang, L., Lewis, C., Phillipson, C., & Yarker, S. (2020). Covid-19 and inequality: Developing an age-friendly strategy for recovery in low income communities [Article]. *Quality in Ageing and Older Adults, 21*(4), 271–279. https://doi.org/10.1108/QAOA-09-2020-0044

Calo, D. (2022). Additional dwelling units: A potential solution to the affordable and senior housing crisis [Article]. *Journal of Urban Regeneration and Renewal, 15*(4), 371–378.

Carpagnano, G. E., Migliore, G., Grasso, S., Procacci, V., Resta, E., Panza, F., & Resta, O. (2021). More skilled clinical management of COVID-19 patients modified mortality in an intermediate respiratory intensive care unit in Italy. *Respiratory Research, 22*(1), 16. https://doi.org/10.1186/s12931-021-01613-2

Chau, H. W., Gilzean, I., Jamei, E., Palmer, L., Preece, T., & Quirke, M. (2022). Comparative analysis of 20-minute neighbourhood policies and practices in Melbourne and Scotland [Article]. *Urban Planning, 7*(4), 13–24. https://doi.org/10.17645/up.v7i4.5668

Chen, J. J., Liu, L. F., Lin, C. I., & Lin, H. C. (2022a). Multidimensional determinants of well-being among community-dwelling older adults during the early stage of the COVID-19 pandemic in Taiwan [Article]. *Gerontology and Geriatric Medicine, 8*. https://doi.org/10.1177/23337214221111227

Chen, X., Wang, H., Ai, J., Shen, L., Lin, K., Yuan, G., et al. (2022b). Identification of CKD, bedridden history and cancer as higher-risk comorbidities and their impact on prognosis of hospitalized Omicron patients: A multi-centre cohort study. *Emerg Microbes Infect, 11*(1), 2501–2509. https://doi.org/10.1080/22221751.2022.2122581

Dabelko-Schoeny, H., White, K., Sheldon, M., Park, C., Happel, C., & Purvis, T. (2022). Age-friendly communities during the time of COVID-19: A model for rapid community response [Note]. *Journal of Aging and Social Policy, 34*(2), 275–292. https://doi.org/10.1080/08959420.2022.2049576

Dejvajara, D., Aungkasuraphan, R., Palee, P., Piankusol, C., Sirikul, W., & Siviroj, P. (2022). Effects of home-based nine-square step exercises for fall prevention in Thai community-dwelling older adults during a COVID-19 lockdown: A pilot randomized controlled study. *International Journal of Environmental Research and Public Health, 19*(17). https://doi.org/10.3390/ijerph191710514

DeLange Martinez, P., Nakayama, C., & Young, H. M. (2020). Age-friendly cities during a global pandemic [Article]. *Journal of Gerontological Nursing, 46*(12), 7–13. https://doi.org/10.3928/00989134-20201106-02
Di Castelnuovo, A., Costanzo, S., Antinori, A., Berselli, N., Blandi, L., Bonaccio, M., et al. (2021). Heparin in COVID-19 patients is associated with reduced in-hospital mortality: The multicenter Italian CORIST study. *Thrombosis and Haemostasis, 121*(8), 1054–1065. https://doi.org/10.1055/a-1347-6070
Dürst, A. V., Graf, C. E., Ruggiero, C., Zekry, D., Boccardi, V., Monney, L., et al. (2022). Fighting social isolation in times of pandemic COVID-19: The role of video calls for older hospitalized patients. *Aging Clinical and Experimental Research, 34*(9), 2245–2253. https://doi.org/10.1007/s40520-022-02188-8
Emerson, K. G. (2020). Coping with being cooped up: Social distancing during COVID-19 among 60+ in the United States [Article]. *Revista Panamericana de Salud Publica/Pan American Journal of Public Health, 44*, Article e81. https://doi.org/10.26633/RPSP.2020.81
Esposti, L. D., Buda, S., Nappi, C., Paoli, D., & Perrone, V. (2020). Implications of covid-19 infection on medication adherence with chronic therapies in Italy: A proposed observational investigation by the fail-to-refill project [Note]. *Risk Management and Healthcare Policy, 13*, 3179–3185. https://doi.org/10.2147/RMHP.S265264
Fang, M. L., Sixsmith, J., Hamilton-Pryde, A., Rogowsky, R., Scrutton, P., Pengelly, R., et al. (2022). Co-creating inclusive spaces and places: Towards an intergenerational and age-friendly living ecosystem. *Frontiers in Public Health, 10*, 996520. https://doi.org/10.3389/fpubh.2022.996520
Figurska, M., Dawidowicz, A., & Zysk, E. (2022). Voronoi diagrams for senior-friendly cities [Article]. *International Journal of Environmental Research and Public Health, 19*(12), Article 7447. https://doi.org/10.3390/ijerph19127447
Finlay, J. M., Meltzer, G., Cannon, M., & Kobayashi, L. C. (2022). Aging in place during a pandemic: Neighborhood engagement and environments since the COVID-19 pandemic onset. *Gerontologist, 62*(4), 504–518. https://doi.org/10.1093/geront/gnab169
Fournier, H., Katsuragawa, K., Singh, K., Bin Hannan, N., & Kondratova, I. (2021a). Designing towards an application to find a nurse. In *International Conference on Human-Computer Interaction*, vol. 1419, pp. 257–264.
Fournier, H., Kondratova, I., & Katsuragawa, K. (2021b). Smart technologies and Internet of Things designed for aging in place. In *HCI for cybersecurity, privacy and trust*, vol. 12788 LNCS, pp. 158–176.
Frantál, B., Klapka, P., & Nováková, E. (2020). When home becomes a cage: Daily activities, space-time constraints, isolation and the loneliness of older adults in urban environments [Article]. *Moravian Geographical Reports, 28*(4), 322–337. https://doi.org/10.2478/mgr-2020-0024
Funk, L., Krawczyk, M., Cherba, M., Cohen, S. R., Dujela, C., Nichols, C., & Stajduhar, K. (2023). 'The beauty and the less beautiful': Exploring the meanings of dying at 'home' among community and practitioner representatives and advocates across Canada. *Palliat Care Soc Pract, 17*, 26323524231156944. https://doi.org/10.1177/26323524231156944
Gaglione, F., Gargiulo, C., Zucaro, F., & Cottrill, C. (2021). 15-minute neighbourhood accessibility: A comparison between Naples and London [Article]. *European Transport – Trasporti Europei*, (85). https://doi.org/10.48295/ET.2021.85.5
Gaglione, F., Gargiulo, C., Zucaro, F., & Cottrill, C. (2022). *Urban accessibility in a 15-minute city: A measure in the city of Naples, Italy*.
Garcia, M. A., Homan, P. A., García, C., & Brown, T. H. (2021). The color of COVID-19: Structural racism and the disproportionate impact of the pandemic on older black and Latinx adults [Article]. *Journals of Gerontology – Series B Psychological Sciences and Social Sciences, 76*(3), E75–E80. https://doi.org/10.1093/geronb/gbaa114
Ghosh, M. (2022). Ageing and technology – Systematic review of the merger of ageing & technology in the new normal era. In *International conference on human-computer interaction*, vol. 1580 CCIS, pp. 516–523.

Golubev, A. G., & Sidorenko, A. V. (2020). Theory and practice of aging upon COVID-19 pandemic [Article]. *Advances in gerontology = Uspekhi gerontologii, 33*(2), 397–408.

Greenwood-Hickman, M. A., Dahlquist, J., Cooper, J., Holden, E., McClure, J. B., Mettert, K. D., et al. (2021). "They're going to zoom it": A qualitative investigation of impacts and coping strategies during the COVID-19 pandemic among older adults. *Frontiers in Public Health, 9*, 679976. https://doi.org/10.3389/fpubh.2021.679976

Greer, V., Canham, S. L., Hong, A., Agutter, J., Garcia Zambrana, I., & Van Natter, J. M. (2023). Aging in place through the COVID-19 pandemic: Perspectives from aging service providers. *Journal of Applied Gerontology, 42*(7), 1530–1540. https://doi.org/10.1177/07334648231159375

Gupta, P., Ibrahim, O., Skubic, M., & Scott, G. J. (2021, 2021). Leveraging unsupervised machine learning to discover patterns in linguistic health summaries for eldercare. *Annual International Conference of the IEEE Engineering in Medicine & Biology Society*, 2180–2185.

Herron, R. V., Newall, N. E. G., Lawrence, B. C., Ramsey, D., Waddell, C. M., & Dauphinais, J. (2021). Conversations in times of isolation: Exploring rural-dwelling older adults' experiences of isolation and loneliness during the covid-19 pandemic in Manitoba, Canada [Article]. *International Journal of Environmental Research and Public Health, 18*(6), 1–15, Article 3028. https://doi.org/10.3390/ijerph18063028

Herron, R. V., Lawrence, B. C., Newall, N. E. G., Ramsey, D., Waddell-Henowitch, C. M., & Dauphinais, J. (2022). Rural older adults' resilience in the context of COVID-19 [Article]. *Social Science and Medicine, 306*, Article 115153. https://doi.org/10.1016/j.socscimed.2022.115153

Hill, J. R., Harrington, A. B., Adeoye, P., Campbell, N. L., & Holden, R. J. (2021). Going remote-demonstration and evaluation of remote technology delivery and usability assessment with older adults: Survey study. *JMIR mHealth and uHealth, 9*(3), e26702. https://doi.org/10.2196/26702

Hoang, P., Whaley, C., Thompson, K., Ho, V., Rehman, U., Boluk, K., & Grindrod, K. A. (2021). Evaluation of an intergenerational and technological intervention for loneliness: Protocol for a feasibility randomized controlled trial [Article]. *JMIR Research Protocols, 10*(2), Article e23767. https://doi.org/10.2196/23767

Hörder, H., Johansson, L., Guo, X., Grimby, G., Kern, S., Östling, S., & Skoog, I. (2018). Midlife cardiovascular fitness and dementia: A 44-year longitudinal population study in women. *Neurology, 90*(15), e1298–e1305. https://doi.org/10.1212/WNL.0000000000005290

Hsu, C. H., Lin, H. H., Wang, C. C., & Jhang, S. (2020). How to defend covid-19 in Taiwan? Talk about people's disease awareness, attitudes, behaviors and the impact of physical and mental health [Article]. *International Journal of Environmental Research and Public Health, 17*(13), 1–23, Article 4694. https://doi.org/10.3390/ijerph17134694

James, B. L. (2023). What housing characteristics support seniors? Seniors' experiences of housing and home in New Zealand during the COVID-19 pandemic [Article]. *International Journal of Housing Markets and Analysis, 16*(3), 552–574. https://doi.org/10.1108/IJHMA-08-2022-0124

Jen, S., Jeong, M., Kang, H., & Riquino, M. (2021). Ageism in COVID-related newspaper coverage: The first month of a pandemic [Article]. *Journals of Gerontology – Series B Psychological Sciences and Social Sciences, 76*(9), 1904–1912. https://doi.org/10.1093/geronb/gbab102

Joseph, C. A., O'Shea, B. Q., Eastman, M. R., Finlay, J. M., & Kobayashi, L. C. (2022). Physical isolation and mental health among older US adults during the COVID-19 pandemic: Longitudinal findings from the COVID-19 Coping Study [Article]. *Social Psychiatry and Psychiatric Epidemiology, 57*(6), 1273–1282. https://doi.org/10.1007/s00127-022-02248-4

Khavarian-Garmsir, A. R., Sharifi, A., & Sadeghi, A. (2023). The 15-minute city: Urban planning and design efforts toward creating sustainable neighborhoods. *Cities, 132*, 104101. https://doi.org/10.1016/j.cities.2022.104101

Kohn, J. N., Lobo, J. D., Troyer, E. A., Wilson, K. L., Ang, G., Walker, A. L., et al. (2023). Tai chi or health education for older adults with hypertension: Effects on mental health and psychological resilience to COVID-19. *Aging & Mental Health, 27*(3), 496–504. https://doi.org/10.1080/13607863.2022.2053836

Kordova, S., & Keisari, S. (2020). 'Great red anemone and its beautiful black pollens': On tele-drama therapy sessions with older adults in times of COVID-19 [Article]. *Drama Therapy Review, 6*, 15–19. https://doi.org/10.1386/dtr_00036_1

Kotwal, A. A., Holt-Lunstad, J., Newmark, R. L., Cenzer, I., Smith, A. K., Covinsky, K. E., et al. (2021). Social isolation and loneliness among San Francisco Bay Area older adults during the COVID-19 shelter-in-place orders. *Journal of the American Geriatrics Society, 69*(1), 20–29. https://doi.org/10.1111/jgs.16865

Krendl, A. C., & Perry, B. L. (2021). The impact of sheltering in place during the COVID-19 pandemic on older adults' social and mental well-being [Article]. *Journals of Gerontology – Series B Psychological Sciences and Social Sciences, 76*(2), E53–E58. https://doi.org/10.1093/geronb/gbaa110

Lennon, J. C., Hantke, N., Mattek, N., Wu, C. Y., Dodge, H., Wall, R., et al. (2022). Mood and activity changes during the COVID-19 pandemic in rural and urban Veterans and their cohabitants [Article]. *Clinical Gerontologist*. https://doi.org/10.1080/07317115.2022.2130123

Liu, L. F., & Lu, C. P. (2022). Community innovation and regional sustainability: A case study on an elder social survey conducted by National Cheng Kung University [Article]. *Journal of Nursing, 69*(3), 19–24. https://doi.org/10.6224/JN.202206_69(3).04

Lõhmus, M., Stenfors, C. U. D., Lind, T., Lauber, A., & Georgelis, A. (2021). Mental health, greenness, and nature related behaviors in the adult population of Stockholm county during covid-19-related restrictions [Article]. *International Journal of Environmental Research and Public Health, 18*(6), Article 3303. https://doi.org/10.3390/ijerph18063303

Manthorpe, J., Woolham, J., Steils, N., Stevens, M., Martineau, S., Owen, J., & Tinelli, M. (2022). Experiences of adult social work addressing self-neglect during the Covid-19 pandemic [Article]. *Journal of Social Work, 22*(5), 1227–1240. https://doi.org/10.1177/14680173221083446

Marston, H. R., Shore, L., & White, P. J. (2020). How does a (Smart) age-friendly ecosystem look in a post-pandemic society? [Article]. *International Journal of Environmental Research and Public Health, 17*(21), 1–43, Article 8276. https://doi.org/10.3390/ijerph17218276

Masters, J. L., Wilkinson, L. R., Kelly, C. M., McKelvey, M., Ghazi Saidi, L., Hill, T. L., et al. (2022). COVID-19 and the vulnerabilities of community-dwelling other adults: Findings from a statewide survey of home-delivered meals recipients* [Article]. *Gerontology and Geriatric Medicine, 8*. https://doi.org/10.1177/23337214221086465

McDonald, A. P., Rizzotti, R., Rivera, J. M., D'Arcy, R. C. N., Park, G., & Song, X. (2021). Toward improved homecare of frail older adults: A focus group study synthesizing patient and caregiver perspectives [Article]. *Aging Medicine, 4*(1), 4–11. https://doi.org/10.1002/agm2.12144

Meuser, T., Cohen Konrad, S., Robnett, R., & Brooks, F. (2022). Telecollaboration in gerontology service learning: Addressing isolation & loneliness in a pandemic [Article]. *Gerontology and Geriatrics Education, 43*(1), 18–33. https://doi.org/10.1080/02701960.2021.1956489

Mikal, J. P., Wurtz, R., & Grande, S. W. (2021). Older adults' computer-mediated communication (CMC) engagement following COVID-19 and its impact on access to community, information, and resource exchange: A longitudinal, qualitative study [Article]. *Gerontology and Geriatric Medicine, 7*. https://doi.org/10.1177/23337214211052201

Miller, J., & McDaniel, T. (2021). Social robotics to address isolation and depression among the aging during and after COVID-19. In *International conference on human-computer interaction*, vol. 1420, pp. 164–171.

Muheidat, F., & Tawalbeh, L. A. (2020). In-home floor based sensor system-smart carpet- to facilitate healthy aging in place (AIP) [Article]. *IEEE Access, 8*, 178627–178638. https://doi.org/10.1109/ACCESS.2020.3027535

Nagajayanthi, B., Vikneshrajan, M., & Navya, K. (2021). User authenticated smart home automation system with healthcare. *Journal of Physics: Conference Series, 2115*, 012031.

Naruse, T., Hatsushi, M., Tsuchiya-Ito, R., Kobayashi, M., & Kato, J. (2023). Experiences of disabled older adults in urban area adult day care centers: A multisite case study [Article]. *Gerontology and Geriatric Medicine, 9*. https://doi.org/10.1177/23337214231156304

Nurain, N., Caldeira, C., & Connelly, K. (2021). Older adults' experiences of autonomy during COVID-19 pandemic. In *Older adults' experiences of autonomy during COVID-19 pandemic*. https://doi.org/10.1145/3411763.3451674.
O'Shea, B. Q., Finlay, J. M., Kler, J., Joseph, C. A., & Kobayashi, L. C. (2021). Loneliness among US adults aged ≥55 early in the COVID-19 pandemic: Findings from the COVID-19 coping study [Article]. *Public Health Reports, 136*(6), 754–764. https://doi.org/10.1177/00333549211029965
Okoye, S. M., Mulcahy, J. F., Fabius, C. D., Burgdorf, J. G., & Wolff, J. L. (2021). Neighborhood broadband and use of telehealth among older adults: Cross-sectional study of national survey data linked with census data [Article]. *Journal of Medical Internet Research, 23*(6), Article e26242. https://doi.org/10.2196/26242
Park, M., & Go, Y. (2021). Dissemination of person-centered community care for mild dementia using local cooperatives in Korea during the COVID-19 pandemic [Article]. *Alzheimer's & Dementia: The Journal of the Alzheimer's Association, 17*, e052873. https://doi.org/10.1002/alz.052873
Pech, M., Gbessemehlan, A., Dupuy, L., Sauzéon, H., Lafitte, S., Bachelet, P., et al. (2022). Lessons learned from the SoBeezy program for older adults during the COVID-19 pandemic: Experimentation and evaluation [Article]. *JMIR Formative Research, 6*(11), Article e39185. https://doi.org/10.2196/39185
Perdana, A., & Mokhtar, I. A. (2022). Seniors' adoption of digital devices and virtual event platforms in Singapore during Covid-19 [Article]. *Technology in Society, 68*, Article 101817. https://doi.org/10.1016/j.techsoc.2021.101817
Perotti, L., & Strutz, N. (2022). Evaluation and intention to use the interactive robotic kitchen system AuRorA in older adults [Article]. *Zeitschrift fur Gerontologie und Geriatrie*. https://doi.org/10.1007/s00391-022-02105-8
Pestine-Stevens, A., & Greenfield, E. A. (2020). The need for community practice to support aging in place during COVID-19 [Article]. *Journal of Gerontological Social Work*, 631–634. https://doi.org/10.1080/01634372.2020.1789258
Phillips, J., Walford, N., Hockey, A., & Sparks, L. (2021). Older people, town centres and the revival of the 'high street' [Article]. *Planning Theory and Practice, 22*(1), 11–26. https://doi.org/10.1080/14649357.2021.1875030
Piccoli, M., Tannou, T., Hernandorena, I., & Koeberle, S. (2020). Ethical approach to the issue of confinement of the elderly in the context of the COVID-19 pandemic: Prevention of frailty versus risk of vulnerability [Short Survey]. *Ethics, Medicine and Public Health, 14*, Article 100539. https://doi.org/10.1016/j.jemep.2020.100539
Pinazo-Hernandis, S., Blanco-Molina, M., & Ortega-Moreno, R. (2022). Aging in place: Connections, relationships, social participation and social support in the face of crisis situations [Article]. *International Journal of Environmental Research and Public Health, 19*(24), Article 16623. https://doi.org/10.3390/ijerph192416623
Prasad, A., Immel, M., Fisher, A., Hale, T. M., Jethwani, K., Centi, A. J., et al. (2022). Understanding the role of virtual outreach and programming for LGBT individuals in later life [Article]. *Journal of Gerontological Social Work, 65*(7), 766–781. https://doi.org/10.1080/01634372.2022.2032526
Principi, A., Lucantoni, D., Quattrini, S., Di Rosa, M., & Socci, M. (2022). Changes in volunteering of older adults in the time of the COVID-19 pandemic: The role of motivations [Article]. *International Journal of Environmental Research and Public Health, 19*(22), Article 14755. https://doi.org/10.3390/ijerph192214755
Rantanen, T., Eronen, J., Kauppinen, M., Kokko, K., Sanaslahti, S., Kajan, N., & Portegijs, E. (2021). Life-space mobility and active aging as factors underlying quality of life among older people before and during COVID-19 lockdown in Finland – A longitudinal study [Article]. *Journals of Gerontology – Series A Biological Sciences and Medical Sciences, 76*(3), E60–E67. https://doi.org/10.1093/gerona/glaa274

Resendiz-Villasenor, C., Rahimian, F. P., Carnemolla, P., Rodriguez, S., & Dawood, N. (2021). Design for ageing in place in isolated contexts: A methods literature review. In *International conference on human-computer interaction*, vol. 12786 LNCS, pp. 74–87.

Reynolds, L. (2020). The COVID-19 pandemic exposes limited understanding of ageism [Article]. *Journal of Aging and Social Policy, 32*(4–5), 499–505. https://doi.org/10.1080/08959420.2020.1772003

Richards, O. K., Marcu, G., & Brewer, R. N. (2021). *Hugs, bible study, and speakeasies: Designing for older adults' multimodal connectedness*. https://doi.org/10.1145/3461778.3462075.

Roe, J., Mondschein, A., Neale, C., Barnes, L., Boukhechba, M., & Lopez, S. (2020). The urban built environment, walking and mental health outcomes among older adults: A pilot study. *Frontiers in Public Health, 8*, 575946. https://doi.org/10.3389/fpubh.2020.575946

Rosa-Jiménez, C., & Nebot, N. (2021). Economic viability of neighborhood cooperatives for active aging. Malaga as a case study [Article]. *Architecture, City and Environment, 15*(45), 1–13, Article 10368. https://doi.org/10.5821/ace.15.45.10368

Rovio, S., Kåreholt, I., Helkala, E. L., Viitanen, M., Winblad, B., Tuomilehto, J., et al. (2005). Leisure-time physical activity at midlife and the risk of dementia and Alzheimer's disease. *Lancet Neurology, 4*(11), 705–711. https://doi.org/10.1016/S1474-4422(05)70198-8

Sekhon, H., Sekhon, K., Launay, C., Afililo, M., Innocente, N., Vahia, I., et al. (2021). Telemedicine and the rural dementia population: A systematic review [Review]. *Maturitas, 143*, 105–114. https://doi.org/10.1016/j.maturitas.2020.09.001

Shang, J., Chastain, A. M., Perera, U. G. E., Quigley, D. D., Fu, C. J., Dick, A. W., et al. (2020). COVID-19 preparedness in US home health care agencies [Article]. *Journal of the American Medical Directors Association, 21*(7), 924–927. https://doi.org/10.1016/j.jamda.2020.06.002

Shao, D., & Lee, I. J. (2020). Acceptance and influencing factors of social virtual reality in the urban elderly [Article]. *Sustainability (Switzerland), 12*(22), 1–19, Article 9345. https://doi.org/10.3390/su12229345

Smith, M. L., Steinman, L. E., & Casey, E. A. (2020). Combatting social isolation among older adults in a time of physical distancing: The COVID-19 social connectivity paradox [Article]. *Frontiers in Public Health, 8*, Article 403. https://doi.org/10.3389/fpubh.2020.00403

Soubutts, E., Ayobi, A., Eardley, R., McNaney, R., Cater, K., & O'Kane, A. A. (2022). Amazon Echo show as a multimodal human-to-human care support tool within self-isolating older UK households [Article]. *Proceedings of the ACM on Human-Computer Interaction, 6*, Article 302. https://doi.org/10.1145/3555193

Spellman, C. W., & Ntiri, D. W. (2022). Helping older African Americans thrive in urban communities: Empowering lessons from Detroit [Article]. *Urban Social Work, 6*(3), 240–255. https://doi.org/10.1891/USW-2022-0003

Stradford, J., Sakhare, A., Ravichandran, R., Schroeder, E. T., Michener, L. A., & Pa, J. (2021). Conducting a VR clinical trial in the era of COVID-19 [Article]. *Frontiers in Virtual Reality, 2*, Article 639478. https://doi.org/10.3389/frvir.2021.639478

Sut, D. J., Kumar, C. V., & Prabhu, S. (2022). *Design and analysis of assistant mobile robot for isolated patients*. Proceedings of the International Conference on Recent Advances in Manufacturing Engineering Research. https://doi.org/10.1063/5.0095624

Szczuka, Z., Abraham, C., Baban, A., Brooks, S., Cipolletta, S., Danso, E., et al. (2021). The trajectory of COVID-19 pandemic and handwashing adherence: Findings from 14 countries. *BMC Public Health, 21*(1), 1791. https://doi.org/10.1186/s12889-021-11822-5

Tai-Seale, M., Cheung, M. W., Kwak, J., Harris, V., Madonis, S., Russell, L., et al. (2023). Unmet needs for food, medicine, and mental health services among vulnerable older adults during the COVID-19 pandemic. *Health Services Research, 58*(Suppl 1), 69–77. https://doi.org/10.1111/1475-6773.14084

Talic, S., Shah, S., Wild, H., Gasevic, D., Maharaj, A., Ademi, Z., et al. (2021). Effectiveness of public health measures in reducing the incidence of covid-19, SARS-CoV-2 transmission, and covid-19 mortality: Systematic review and meta-analysis. *BMJ, 375*, e068302. https://doi.org/10.1136/bmj-2021-068302

Tan, T. H., & Idris, I. (2023). Assessing the significance of first place and online third places in supporting Malaysian seniors' well-being during the pandemic. *Humanit Soc Sci Commun, 10*(1), 149. https://doi.org/10.1057/s41599-023-01655-5

Tiilikainen, E., Lisko, I., Kekkonen, E., Solomon, A., Ngandu, T., Kivipelto, M., & Kulmala, J. (2021). Everyday life meaningfulness for the community-dwelling oldest old during the COVID-19 pandemic [Article]. *Frontiers in Psychology, 12*, Article 716428. https://doi.org/10.3389/fpsyg.2021.716428

Van Hees, S., Greubel, C., Moors, E., & Peine, A. (2021). Valuation in health and ageing innovation practices [Article]. *Ageing and Society*. https://doi.org/10.1017/S0144686X21001483

Vanleerberghe, P., De Witte, N., Claes, C., Schalock, R. L., & Verté, D. (2017). The quality of life of older people aging in place: A literature review. *Quality of Life Research, 26*(11), 2899–2907. https://doi.org/10.1007/s11136-017-1651-0

Washington, S. E., Edwards, E., Stiles, D. L., & West Bruce, S. (2023). Implementation of the CAPABLE program with older adults during the COVID-19 pandemic [Article]. *OTJR Occupation, Participation and Health*. https://doi.org/10.1177/15394492231151885

Yeh, M. H., Huang, C. H., Lin, Y. C., Huang, T. J., & Chen, M. Y. (2022). The health impact of an innovative summer camp for older adults: A pilot study using an interdisciplinary collaborative approach [Article]. *BMC Nursing, 21*(1), Article 4. https://doi.org/10.1186/s12912-021-00782-1

Young, D. R., Hong, B. D., Lo, T., Inzhakova, G., Cohen, D. A., & Sidell, M. A. (2022). The longitudinal associations of physical activity, time spent outdoors in nature and symptoms of depression and anxiety during COVID-19 quarantine and social distancing in the United States. *Preventive Medicine, 154*, 106863. https://doi.org/10.1016/j.ypmed.2021.106863

Zhu, Y., Liu, Y., & Jiang, H. (2022). Geriatric health care during the COVID-19 pandemic: Managing the health crisis [Review]. *Clinical Interventions in Aging, 17*, 1365–1378. https://doi.org/10.2147/CIA.S376519

Zingmark, M., Nordeström, F., & Iwarsson, S. (2022). Challenges related to self-assessment of active ageing during the Covid-19 pandemic in Sweden. *BMC Research Notes, 15*(1), 171. https://doi.org/10.1186/s13104-022-06059-3

Chapter 11
Demographic Changes and Ageing in China and India: A Public Policy Perspective

Jamie P. Halsall, Liping Mei, Kalim Siddiqui, Michael Snowden, and James Stockton

11.1 Introduction

Citizens across the world are living longer due to improvements in living standards and the healthcare system (Cook et al., 2015; Cook & Halsall, 2012; Haque & Afrin, 2022; Karlin & Weil, 2017; Powell, 2023a, b; Rapp et al., 2022). This is no more apparent than in the findings of a recent report by the World Health Organization (WHO) (2022), which points out that the speed and rate of population ageing is far greater than ever before and, from 2015 to 2050, the dimension of the world's population over 60 years of age will approximately double from 12% to 22%. As the WHO (2022) notes, population ageing has created public policy challenges in the economic and social context. Obviously, this will have broader implications for the environment, and the adverse impact of climate change will have implications for the ageing populations in China and India.

In their 2012 work, Cook and Halsall note that when examining ageing from a social science perspective, there are multifaceted problems in local, national, and global contexts. A decade on, Powell's (2023b) research demonstrates that, as ageing populations continue to grow, new debates about the economic and social aspects of ageing are emerging. At the centre of this academic work is the focus on 'ageing in place', and, as recently noted by Barati et al. (2022), this terminology has

J. P. Halsall (✉) · M. Snowden · J. Stockton
School of Human and Health Sciences, University of Huddersfield, Queensgate, UK
e-mail: j.p.halsall@hud.ac.uk; m.a.snowden@hud.ac.uk; James.Stockton@hud.ac.uk

L. Mei
Beijing Youth Politics College, Beijing, China
e-mail: meiliping@bjypc.edu.cn

K. Siddiqui
Huddersfield Business School, University of Huddersfield, Huddersfield, UK
e-mail: K.U.Siddiqui@hud.ac.uk

N. Asiamah et al. (eds.), *Sustainable Neighbourhoods for Ageing in Place*,
https://doi.org/10.1007/978-3-031-41594-4_11

become popular with public policymakers. Ageing in place has been defined as a concept that primarily focuses on ageing individuals (i.e. 65 years old and older) that remain in their local communities with selected levels of independence rather than living in residential care homes. For a long period of time, there has been a general international consensus that 'Having people remain in their homes and communities for as long as possible also avoids the costly option of institutional care and is therefore favoured by policy makers, health providers, and by many older people themselves' (Wiles et al., 2012, p. 357). Hence, the authors of this chapter examine the ways in which China and India approach ageing within a public policy context. This chapter has five sections. It begins with a comparison between China and India to provide comparative points of reference regarding the economic transformation that has occurred in both developing nations. The second section analyses demographic changes that have developed in China and India. Then, from this, the next two sections specifically focus on 'ageing in place' as a public policy and how this is a driver for addressing social and economic pressures facing China and India. The final section will summarize the key content and provide a discussion on the importance of the chapter's findings.

11.2 Economic Transformation in China and India

Both China and India have experienced noteworthy economic development since the introduction of significant economic reforms. Their economies have emerged as important global economic players and now play a crucial role in the contemporary global economy. Economic upturn or downturn in both or either of these economies arouses global concern. Undoubtedly, there is a substantial difference in the economic size of China and India, but the growth rate and the potential for economic growth show some similarities, despite the fundamental differences in their economic policies and political governance (IMF, 2020).

China began economic reforms through a gradual process of economic liberalization in 1978, while India's reforms began in 1991 (Siddiqui, 2018a). A gradual opening up of the economy and withdrawal of trade barriers were the natural offshoots of the policy options hitherto chosen in each country. With the withering away of 'protectionist' policies, India's and China's trade patterns are likely to march in the direction of their comparative advantage. It has also been observed that since economic reforms were undertaken by these two economies, both have recorded high gross domestic product (GDP) growth. Trade between India and China gained momentum since India's 1991 economic policy reform and accelerated further after China acceded to the World Trade Organization (WTO). China's and India's growth performance and trade also received a boost after their integration with the world economy (Siddiqui, 2018b).

The huge populations and market sizes of China and India make it very important to focus on these economies, as their performance would impact the global economy. For instance, China will become the world's largest economy, overtaking the USA by 2030, while India overtakes Japan, Germany, the UK, and France to become number 3. In 2021, China's economy (US$ 17.7 trillion) was more than five times the size of India's (US$ 3.2 trillion). In 2021, China's GDP accounts for 18.5% of the world's total. The inflow of foreign capital has played a very important role in the rapid economic growth in both economies. China's net inflow of foreign direct investment (FDI) (US$ 253) was nearly four times that of India (US$ 64) in 2020 (Parkin & Singh, 2022).

The economic development patterns of two countries differ in many respects. China's sharp rise in growth rates has been accompanied by significant increase in domestic savings and investments, while India too has accompanied growth rates since pro-market reforms were launched with much lower rates of investments. One remarkable difference with China was where during the reform period, the manufacturing share of GDP increased, whereas India witnessed almost stagnant industrial sector share of GDP and was unsuccessful to move surplus low productive workforce from agriculture sector to industrial sector. Moreover, China relied largely on cheap labour and moved them into expanding manufacturing. However, India's growth was accompanied by almost no employment creation, which is also known as job-less growth. The growth in India more relied on human capital-intensive industries to induce its growth (Parkin & Singh, 2022).

Indeed, high growth rates were not accompanied by transfer of labour from agriculture to industrial sector, and problems of underdevelopment in rural areas remain largely unsolved in India. An extraordinary high investment share in GDP has played an important role in China's rapid economic transformation – a sharp contrast to India where growth has not sustained over several decades and concentrated in few regions of high skilled and high productivity activities, especially in services (Siddiqui, 2020b).

China has achieved significant transformation in its economy over the last four decades, which has been debated by the academics whether it would lead to decline of the Western economic and political domination. Despite the slow growth since the COVID-19, Chinese economy grew on average 6.6% per annum, which was higher than world's average growth of 2.6% between 2013 and 2021. During the same period, China's contribution to global economic growth on average was 38.6%, which was higher than G-7 countries. Moreover, China in 2020 emerged as the world's largest trading nation for the first time, and in 2021, its foreign trade reached to US$ 6.9 trillion (The Economist, 2023).

Furthermore, China's per capita income grew steadily reaching US$ 11,890 in 2021, doubling from 2012. The important domestic factors included a sharp increase in spending on infra-structure, healthcare, and education; as a result, the average life expectancy of Chinese people reached 79.9 years in 2020, 5.2 years above the global average. However, the COVID-19 pandemic hit hard, and growth fell sharply, but soon after in first quarter of the 2021, its economy grew at 18.3%. It is widely believed that China, rather than the USA, would be the major driving force to take the global economy out of the current recession (The Financial Times, 2022; Siddiqui, 2020b).

In 1978, Chinese government introduced economic reforms under the leadership of Deng Xiaoping on a limited scale to attract foreign capital inflows, which had very positive impact in increasing output and exports. Moreover, by joining the World Trade Organization (WTO) in 2001, China began successfully to deepen its relations with global corporations and value chains. The opening of global markets for China led to enormous increase in trade. And because of various polices, China has successfully emerged from low-income to a middle-income country. In the late 1970s, when China embarked on pro-market reforms, the percentage share of agriculture in total employment was about the same as India, while the percentage of value added was much higher in India compared to China. For the same period, per capita GDP in purchasing power parity (PPP) terms in India was higher than China. In China, the share of industry was higher than in India, also in terms of value added and employment. However, the share of services was lower in China compared to India in 1978, while the agriculture value added was higher in China. This also meant that the ratio between productivity in industry and agriculture was higher in China than in India.

Globalization and integration with the global economy has certainly benefited China enormously in several ways. After decades of endemic poverty, hundreds of millions of Chinese people have witnessed a net increase in their incomes and improvements in their living conditions, and after paying of necessities, they are able to save or spend on modern consumer products such as trendy clothing or high-tech gadgets.

Between 1992 and 2018, China continued to experience higher growth rates than India; however, both economies had seen acceleration in growth rates. Moreover, both countries witnessed a sharp rise in their exports, capital accumulation, and inflows of foreign capital into their economies for the same period. China was able to rapidly reduce the percentage of agriculture, both in terms of employment and in value added. China's manufacturing sector increased its absolute and relative size much larger than India. Overall, China's industrial sector grew much larger in size, especially in the production of machines, steel, textiles, and automobiles, while India's services grew much faster, especially software services and pharmaceuticals (Siddiqui, 2020a).

China's savings and investment were much higher compared to India's after the economic reforms were introduced. China opened its economy to foreign investment and trade nearly 14 years earlier than India. The inflow of foreign capital stock as a percentage of GDP was US$ 5.1 billion in China in 1990, which sharply rose to US$ 16.2 billion in 2000. However, due to the global financial crisis, FDI declined to US$ 9.9 billion and rose again to US$ 11.5 in 2015. India had also witnessed a sharp rise in capital inflows from just merely US$ 0.5 billion in 1990 to US$ 3.5 billion in 2000, US$ 12.3 in 2010, and US$ 12.6 billion in 2015 (The Economist, 2023).

China and India are the two emerging economies in the world. As of 2021, China and India are the second and fifth largest economies in the world, respectively, on a nominal basis (see Table 11.1). On a PPP basis, China is first, and India is in third place. Both countries share 21% and 26% of the total global wealth in nominal and

Table 11.1 China and India's macroeconomic indicators in 2022

Variable	China	India
Population	1.43bn	1.43bn
GDP growth (%)	4.7	5.1
GDP per head (US$)	13,680 (PPP: 23,460)	2430 (PPP: 8810)
Inflation (%)	2.9	5.2
Budget balance (% of GDP)	−4.2	−6.0

Adapted from: The Economist (2023), p. 103

PPP terms, respectively. Among Asian countries, China and India together contribute more than half of Asia's GDP.

According to statistics, in 2018, the sector wise GDP composition of China was agriculture (8.3%), industry (39.5%), and services (52.2%), while India's was agriculture (15.4%), industry (23%), and services (61.5%).

After independence in 1949, China adopted a socialist economy, both in agriculture and industry, and the private sector was minimal. In China, the Communist Party had retained monopoly power, but after 1978 the economic liberalization market mechanism became the major allocator of resources. And, nearly 95% of consumer prices are still determined by market forces, though the government still control prices in a few important sectors such as finance, power, utilities, and telecommunications.

In China, foreign investments were in a joint venture with state owned enterprises in the 1980s, but most capital has been invested in wholly owned or joint-venture private companies since 1995. Since the early 2000s, business owners have been allowed to be members of the Communist Party, and by 2010 there were more than 12 million such members (Bardhan, 2020). Furthermore, in 2020, nearly one-third of private entrepreneurs were members of the Communist Party, i.e. *xiahai* entrepreneurs.

China is a development state, where the role of public-owned banks, e.g. China Development Bank, has been crucial to provide funds to industries. It seems that the financial system was there to facilitate state directed industrial policy. As Pranab Bardhan argues: 'Successful private companies in China, such as Huawei and Lenovo, have benefitted a great deal from their close ties with the government'. 'Like in the rest of East Asia export promotion combined with domestic technological capacity building and state encouragement of trial and experimentation in exploring comparative advantage sometimes at the expense of static allocation efficiency have been at the core of the development strategy' (2010, p. 247). In India between 1950 and 1980, the key and strategic industries were in the public sector, but India had a much larger private sector and is co-existing with private sector in the same industry such as steel, power plant, and airlines. In India, the informal sector was much larger than in China and employs nearly 94% of the labour force and contributes nearly half of the non-farm output.

Since the early 1990s, in the wake of neoliberal reforms that began with higher economic growth, the service sector became both dominant and the fastest growing sector of India's economy. Between 1992 and 2019, India's economy grew at an average rate of 7% annually. The growth was largely led by services, which grew at the rate of 8% per year. By 1993, services took over as the dominant sector in India's economy, and its share in aggregate gross value added (GVA) stood at 41%, which was larger than agriculture (34%) and in fact much larger than manufacturing (15%). Just over a quarter of a century later in 2019, the share of services in total value added rose to 53%, while the agriculture sector declined to 15%, and manufacturing only saw a small increment to 18%. In India, services have clearly been the more intensive sector relative to other sectors.

However, in 2020, the Indian economy suffered a major setback due to the COVID-19 pandemic and the related lockdowns that were imposed to control it, which caused a sharp economic decline. This also led to a substantial loss of jobs, which greatly affected low-skilled workers from low-income groups in manufacturing. The 2 years of the pandemic's greatest impact served to intensify the trends of exclusive growth, growing income inequality, increasing skill intensity of jobs, and slow exclusion of low skilled from job opportunities (Siddiqui, 2020b).

The Indian economy witnessed sharp rise in growth rates, but that has largely benefitted the wealthy. The richest 10% of India's population has received the largest share of the increased income generated by growth. Moreover, the number of available jobs in high skilled services has been rising disproportionately, while the demand for low-skilled jobs has progressively reduced. It seems that the sources of India's exclusive growth lie in the nature and characteristics of the lead sectors, namely, skill intensive services.

Private businesses have a long history in India. After independence, the state role became more crucial, especially in leading the heavy and strategic industries and the business environment in a heavily regulated industrial environment. Starting-up businesses required permits and licences in exchange for bribes and political funding for elections developed between big business enterprises and political elites (Siddiqui, 2016).

According to the International Monetary Fund (IMF), China's economic expansion will have global impact, as China is the biggest contributor to global GDP and is set to account for more than one-fifth of the total global growth in the next 5 years. There are several factors exerting drags on the economy, namely, the sluggishness in the property market, which is estimated to be contributing to nearly 30% of the GDP and which stems from the government's insistence that real estate must reduce debt ratios. The other major impediment to growth is 'zero-COVID' strategy (The Financial Times, 2022).

Soon after independence, India failed to implement radical land reforms and confront feudal forces head on, which curtailed the possibility of creating rural equality. Now, seven decades after independence, the unfinished task of unshackling the agrarian structure has led to the perpetuation of an exploitative system in the countryside, marked by mass poverty, malnutrition, disease, and deprivation. After

Table 11.2 India's macroeconomic variables from 2016 to 2021

Year	Private consumption expenditure (%)	Private investment expenditure (%)	Government spending (% of GDP)	Fiscal deficit (% of GDP)	Real GDP growth (%)
2016–2017	8.1	21.27	12.8	3.5	8.3
2017–2018	6.2	21.46	12.53	3.5	7
2018–2019	7.6	22.11	12.2	3.4	6.1
2019–2020	5.5	21.82	13.21	4.6	4.2
2020–2021	−9	−16	17.6	9.5	−7.3

Adapted from: Reserve Bank of India (2021)

the impetus given to agricultural production with the launching of the green revolution in the mid-1960s and 1970s, the failure to implement agrarian reforms, and current neoliberal policies, has led to a prolonged agrarian crisis (Siddiqui, 2015).

In India, 90% of workers were employed in agriculture, industry, and services (Reserve Bank of India, 2021). Many rural workers showed the vulnerability of employment in rural areas during the pandemic period, as more than 70% of India's population still live in rural areas. The demand for work under the government rural work guarantee scheme (MGNREGS) rose suddenly, but there was not enough work available to these households.

In India, the agriculture sector is the largest of the unorganized sectors and employs about 42% of the total workforce in the country. The sector has been suffering for some time, from a decline in public and private investment and a decline due to non-remunerative prices for its commodities; the latter was the reason for big protests by the farmers in 2020 (Siddiqui, 2021).

According to data, the country has been facing demand deficiency since 2017, along with falling consumption, investment, and stagnation in government spending, as shown in Table 11.2. The labour reforms and high employment have squeezed in real wages, resulting in rising profits whose marginal propensity to consume is much less than that of workers. Capitalists will hesitate to invest if they do not expect the aggregate consumers' demand to rise.

11.3 Demographic Changes in China and India

The United Nations Economic and Social Affairs report (2023) indicates that India has surpassed China for total population, highlighting discussions surrounding global demographics and changes. This section explores the demographic changes that underpin the population shifts, and while these may seem to happen naturally, they are driven by historical and long-standing policies derived from population ethics (Budolfson & Spears, 2021). Therefore, when examining demographic changes, consideration for how these historical policies may have affected contemporary statistics should be considered. In the case of the countries examined in this chapter, the starting point for policy consideration is the years immediately following World War II, followed by an examination of the changes occurring in the 1970s through the 1980s and, finally, the early to mid-2010s. When examining these periods, we begin by discussing India, before moving onto a discussion of China, ending with a discussion of how historical demographic changes have impacted modern demographics within both countries.

11.3.1 Historical Policies: India

At midnight on 14 August 1947, India officially achieved independence from the British Empire, ending nearly two centuries of British rule (Robinson, 2019). While India was now independent, there were, and remain, many internal divisions. The policy focused on in this section is the 1952 Family Planning Policy (FPP). The FPP's aim was to control the fertility rate (5.9 at time of implementation) to curb India's projected population growth. The FPP attempted this primarily through the provision of birth control (Lalchandani et al., 2021). In the decade that followed the introduction of the FPP, the fertility rate would decrease by only 0.20–0.30, and it would not be until 1989 that fertility rates would see a significant drop from an average of six children per woman to four (Macrotrends, 2023a).

While the 1990s are commonly considered the start of India's economic boom, it was during the 1980s that the seeds of this were planted. Following the assassination of Indira Gandhi, her successor Rajiv Gandhi implemented a series of policy changes, focusing on expanding India's telecommunication industry and abolishing some licence-inspector raj policies. While this experienced little immediate success, they provided his future successor, Pamulaparthi Venkata Narasimha Rao, with the platform to modernize and strengthen India's economy. This economic upturn was vital, as by 1991 India was approaching bankruptcy (Robinson, 2019). While it would be unfair to claim that this economic upturn caused a direct increase in life expectancy, it correlates with an increase in life expectancy from 55 to 60 years. While an increase of 5 years may not seem significant, in the 27 years that followed, life expectancy only increased by 10 years (Statista, 2023).

During the early to mid-2010s, two significant events occurred; India's total population surpassed 1 billion, and the fertility rate would fall to 2.3 (World Bank, 2023), officially bringing India's fertility rate in line with the global replacement rate standard of 2.1 (World Population Review 1, 2023). 2009 also saw India formally introduce the right to free and compulsory education act, providing aid for the poorest of students to attain education, joining 134 other countries in declaring education a fundamental right of every child (Cheruvalath, 2015).

11.3.2 Historical Policies: China

Like India, post war China saw little decrease in fertility rates, this would drop temporarily in the late 1960s; however, it would not be until the beginning of the 1970s that fertility rates would drop below 6, this decrease occurring over a decade before India's (Macrotrends, 2023b). The stable fertility rate can, be attributed to a period of relative stability following the cessation of World War Two, including the end of the civil wars of the preceding years (Yi et al., 1993). During the post war period, the CCP would prioritize education – the beginning of a recurring focus on improving access to and quality of education throughout the country (Fu & Larbi, 2022).

In the 1970s, China introduced the two-child policy, although it was quickly replaced by the more recognized one child policy at the beginning of the 1980s (Feng et al., 2016). These policies strictly limited the number of children per family, although rare exceptions did exist. The combination of these two policies would see the fertility rate plummet; while these policies were being enforced, the national fertility rate would fall from its peak of 6.3–1.6 (Macrotrends, 2023b). While China reported a fertility rate lower than India during the 1970s, the implementation of the one child policy in the 1980s is where the fertility rate would decrease most rapidly. We can see how dramatic this was by comparing the fertility rates of both countries. In 1985, India reported a fertility rate of 4.5, while China was reporting a fertility rate of 2.6. This 2.6 rate represented a 50% decrease in China's fertility rate since 1975 (Macrotrends, 2023a, b).

In late 2016, China repealed its one child policy, officially allowing families to have two children for the first time in 36 years. Although there are many reasons for this decision, it is likely that the single biggest factor was that the fertility rate had been below the replacement rate standard for several decades. This, combined with an increase in the standard of living, had left China with a large proportion of elderly people and a proportionately small younger population. Other potential issues included the increased burden placed on a single child to support two parents (Feng et al., 2013). Regardless, while the one child policy was repealed, fertility rates did not immediately increase, and in contemporary times, the fertility rate remains below the global replacement level (Macrotrends, 2023b).

As demonstrated, both countries adopted differing policies that came to have wildly differing effects on both population and financial demographics. Key differences included a moderate vs. firm approach to fertility control, differing approaches to educational reform, and variances in the timing of each country's financial boom period.

11.3.3 India and China Contemporary Demographic Discussion

Currently, India and China have the largest populations compared to any other countries in the world, being the first and second most populated countries, respectively (World Population Review, 2023). While India recently surpassed China for population, both countries have, at the time of writing, four times the total population of the USA, the world's third most populated country. Between them, India and China represent 35.66% of the world's total population (World Population Review 2, 2023).

While the countries are roughly congruent in terms of total overall population – with the differences being only in the millions – there are significant differences in terms of their demographics.

One example of these differences can be found in the elderly population. In China, the elderly population is over double that of India: 254 million (Lancet, 2022) compared to 104 million (World Health Organisation, 2023). This discrepancy can be linked to both the large difference in birth rate since the 1970s and the difference in life expectancy between both countries. Life expectancy is higher in China, with an average life expectancy of 78 years (Yao et al., 2023) compared to India's 70.2 years (Statista, 2023).

The difference in elderly population results, inevitably, in differences in youth population and total average age. At the time of writing, India has an average age of 28.7 years, while China's average age is 38.4 years. In terms of population percentages, the under 20s in India make up 34.8% of the population, while China's under 20s make up 23.2% of their population (World Data, 2023).

In terms of fertility rate, India now has a fertility rate of 2.1, while China's fertility rate is 1.7. Recently, China's fertility rate has risen; however, this is minimal; in the last decade, the fertility rate has risen by only 0.06. Currently, India's fertility rate is on the threshold of global replacement standard, while China's is still below (Macrotrends, 2023a, b).

In terms of population density, India has three times the population density when compared to China, with a population density of 464 per km2 compared to 153 per km2 (World Population Review 2, 2023). While both nations have an imbalanced male-to-female ratio, India's population skews more toward males when compared to China's, with a male-to-female ratio of 108/100 (South et al., 2014) compared to China's 104/100 (Tang et al., 2022).

Finally, in terms of education, China is approaching a near 100% secondary education rate, while only 50% of India's children will enter schooling (Jha & Rani, 2016). Estimates put China's population at around 90% literate, compared to India's 60% (Jha & Rani, 2016), although it is worth noting that both China's and India's rates of education are far higher in their urbanized areas than rural ones.

11.4 Ageing in Place: A Case Study of China

China entered the stage of mild population ageing in 2000 and transitioned into a moderately aged society within just over 20 years. According to official data, as of the end of 2022, the population aged 60 and above in the country reached 280.04 million, accounting for 19.8% of the total population, while the population aged 65 and above reached 209.78 million, accounting for 14.9% of the total population.

The significant increase in the elderly population, accompanied by the highly concentrated process of population ageing, presents immense challenges for ageing in place in China (Zhang (2016); Ji (2022).

China has a tradition of 'family-based elderly care' that spans thousands of years. Influenced by traditional filial piety culture, the idea of 'raising children to secure old age' has deeply rooted in the Chinese society and continues to influence the choices people make regarding elderly care. The majority of elderly people are unwilling to opt for nursing home and prefer to remain in their own homes. However, due to industrialization, urbanization, and the impact of family planning policies, the traditional model of 'family-based elderly care' has become unsustainable. The '4-2-1' family structure and long distances make it challenging for most children to care for their ageing parents. Faced with this reality and demand, the Chinese government has gradually established three types of elderly care – home-based care, community-based care, and institutional care – with the emphasis and foundation being placed on the 'community' aspect. According to the 12th Five-Year Plan for the Development of China's Aging Industry in 2011, the structural goals for the three types of elderly care were set at 90:7:3, with 90% of the elderly receiving home-based care services, 7% receiving community-based care services, and 3% receiving institutional care services. This goal has become the main guiding principle for the construction of elderly care facilities across the country and has persisted to this day (Huang Shisong & Sun Shuyan, 2021). Although adjustments have been made based on the specific conditions of population ageing in different regions (such as Shanghai's proposal of 90:6:4), overall, the facility layout follows this principle. Regardless of whether it is '9073' or '9064', it signifies that the 'community' as the elderly's place of residence will bear more than 90% of the elderly ageing in place.

Since the 18th National Congress of the Communist Party of China in 2012, guided by a series of national policies, the integration and development of 'home-based care' and 'community-based care' have been further promoted. A coordinated and comprehensive elderly care service system combining 'home-based care, community-based care, and institutional care' has been gradually established and improved. In practice, the integration of community- and home-based care has led to the formation of the 'community-based home care' model. This model relies on community-based elderly care services and facilities provided by the government and society to meet the elderly's care needs while residing at home. The specific elderly care services include daily living assistance, medical care, emergency response, cultural and recreational activities, spiritual comfort, health management, and rehabilitation support. To facilitate the implementation of the policy of ageing in place, the central government has actively supported pilot reforms of home-based and community-based elderly care services through financial assistance. From November 2016 to the end of 2022, the Ministry of Civil Affairs and the Ministry of Finance have identified 203 pilot areas nationwide. In 2021, the Basic Elderly Care Services Enhancement Action Project was launched in 42 regions (Yang, 2019).

Through trials and explorations, various community-based models of ageing in place have been developed. Some typical examples include Beijing's 'Senior Care

Stations' model, Shanghai's 'Embedded Care' model, and the 'Neighborly Assistance' model in rural areas.

In Beijing's model, care stations are established in communities, serving as the foundation of a top-down 'four-level' (city, district, street or township, community, or village) elderly care service system. This model ensures that elderly care services are delivered precisely to the elderly 'in their vicinity, by their side, and at their bedside', helping them achieve ageing in place. By 2022, Beijing had built over 1000 senior care stations. These care stations typically feature self-therapy equipment areas, daytime care centres, traditional Chinese medicine clinic, counselling rooms, game rooms, and multipurpose activity rooms, providing a variety of services such as healthcare, rehabilitation, disability support, home assistance, psychological counselling, emergency services, and cultural entertainment. In 2022, Beijing further promoted the upgrading of ageing in place services by initiating the construction of street and township elderly care alliances. These alliances coordinate and integrate various elderly care resources within their jurisdiction, including elderly care centres, care stations, medical service institutions, businesses, and public service providers, forming a joint force for ageing in place. The goal is to create a '15-minute living circle' where elderly people can access elderly care services, medical services, and various daily services within a 15-minute walk. In recent years, Beijing has been at the forefront of developing an age-friendly society in China. Many age-friendly neighbourhoods have been achieved by creating a livable environment for the elderly in communities, implementing barrier-free infrastructures, and fostering a social context that values filial piety and respect for the elderly.

Shanghai's 'Embedded Care' service model involves the integration of mini elderly care institutions and comprehensive community service centres embedded within the community. These centres provide multifunctional and comprehensive elderly care services to senior citizens in the surrounding community. Typically, the comprehensive community service centres offer day care, rehabilitation care, convenience stations, recreational activities, dining services, medical treatment, and psychological counselling. They also provide various salons and technology-assisted elderly training courses. Elderly residents can enjoy a wide range of services in these centres (Zhu, 2017). These innovative models demonstrate efforts to promote ageing in place, ensuring that elderly people receive comprehensive and accessible care in their neighbourhoods.

Currently, with 'actively addressing population ageing' becoming a national strategy in China, urban communities across the country have generally established community-based home care service systems that are like those in Beijing and Shanghai. Overall, the policies and practices of ageing in place in urban communities, guided by the national strategy of ageing, focus on social security, elderly care services, health support, and livable environments. They have laid the foundation for the concept and framework of 'elderly-friendly communities' and are striving to create and improve the '15-minute living circle' in communities. The goal is to ensure that elderly people can access the facilities and services necessary for their elderly life within a short distance. However, due to uneven development, there are significant differences in the economic development levels of different regions and cities at different levels. This leads to differences in the quality and level of elderly care and the extent to which the needs of elderly people are met.

Compared to urban areas, elderly care in rural China has always been a challenging issue. The fundamental reason and the biggest challenge for rural elderly care lie in the 'urban-rural dual structure' in China. This dual structure leads to several limitations on ageing in place for rural seniors. Firstly, the pension of rural people is low and far from sufficient to meet their living needs. Secondly, rural medical resources are scarce, with inadequate healthcare services and low healthcare levels, unable to meet the healthcare needs of rural seniors as they age in place. Thirdly, the phenomenon of empty-nest elderly, who live alone, is widespread in rural areas, and there is a lack of cultural and recreational facilities, resulting in a more pronounced problem of social isolation among rural elderly people (Wu, 2019).

Due to the shrinking size of rural families and the large-scale migration of rural youth for work, rural elderly people can no longer rely on their children for care. Consequently, an emerging trend is the shift toward self-care, neighbourhood assistance, and other forms of elderly care. Successful examples of this trend include the 'Mutual Happiness Institute' in Feixiang County, Hebei Province; the 'Joyful Companionship Project' promoted by the China National Committee on Aging; and the 'Women and Elderly Villagers' project implemented by the Hebei Lotus Public Welfare Foundation (Lu, 2022). However, overall, the development of socialized elderly care in rural areas has been slow, and neighbourhood mutual assistance for ageing in place is still in its early stages.

In summary, China has made rapid progress and accumulated significant successful experiences in both policies and practices of ageing in place over the past decade. However, it also faces various challenges. These include the imbalance of urban-rural and regional development and institutional barriers in the implementation of ageing in place policies in communities, particularly the fragmentation of policies and resources due to the involvement of multiple government departments responsible for elderly care. Communities face immense pressure and difficulties in coordinating and managing the diverse responsibilities ('thousands of threads above, but only one needle below'). Other challenges include the mismatch between the spatial distribution of community-based elderly care resources and the scale of the ageing population, sustainability issues in the development of community-based elderly care service providers, low economic income, and the lack of a long-term care insurance system among the elderly. Furthermore, the supply-demand imbalance and inadequate elderly participation caused by one-sided policies and services are also significant challenges. In the context of normalizing epidemic prevention and control, addressing these issues requires the joint efforts of the government, society, communities, families, and individuals. By working together, China can create an environment that enables older people to age in place healthily and with dignity.

11.4.1 Ageing in Place: A Case Study of India

This section of the chapter illustrates the context and challenges presented by the demographic changes within India and explores the impact that an ageing population has on health and wellbeing and its relationship with policy formulation within the notion of ageing in place.

The key longitudinal study of Bloom et al. (2021) drawing upon data provided by the United Nations (2019) illustrates that the population aged 60 or over in India is the world's second largest. In 2019 this was estimated to be 140 million people compared to 250 million people in China. Strikingly the growth rate of the 60 and over age group is growing three times higher than the overall population growth rate of India, consequently resulting in the projection that India's population within the aged 60 and above will form more than 19% in 2050 of the total population, compared to 9% in 2015 (United Nations, 2019). This exponential rise in this societal group presents key challenges for Indian policymakers to design and deliver a response to the public health, social, and economic consequences of this demographic shift within India.

The age-related health burden of disease (Bigonnesse & Chaudhury, 2020) in the forms of the increasing prevalence of cancer, mental health issues, cardiovascular disease, respiratory diseases, diabetes, musculoskeletal issues associated with reduced mobility, and sensory awareness are exacerbated by prevailing demands in health and social care systems (Goel & Kaur, 2022). For example, the necessity to accommodate the dynamic needs of a growing ageing population, this is often thwarted by other priority concerns such as maternal and child health, infectious disease control, noncommunicable diseases, environmental health, air pollution, nutrition and food safety, and tobacco control.

The increasing number of this population group is the result of several key factors: declining fertility rates due to improved access to female education, increased availability and access to contraceptives, and declining infant mortality. Further distinct factors include medical advice, improved access to food and higher nutritional content, and enhanced water supply and sanitation. Reviewing the fertility rate within India over the past 70 years provides further illustration of the challenges faced. The total fertility rate has decreased from 5.9 in 1950 to 2.3 in 2013 and is projected to drop further to 1.88 by 2050, which is below the replacement level (Government of India, 2021; Agarwal et al., 2020). While India is not alone in facing significant economic and social challenges, couple with the changing demographic, the impact has been significant. The changes within the traditional family structure coinciding with the transition from a state-led planned economy to a market economy (Siddiqui, 2018a) have further exacerbated the challenge of ageing in place.

Ageing in place is a term described as living in a chosen and desired community independently rather than in institutional care and widely accepted that people prefer to 'age in place' as it is seen as enabling the older person or 'elder' to maintain independence, autonomy, and connection to social support, including friends and family (Pani-Harreman et al., 2021). This approach to accommodate an ageing population has been adopted internationally as a response to the goal of helping people to remain in their own homes so that they can retain social connections and the support of family. However, the places in which people grow old can be hostile, challenging and disabling without appropriate assessment and planning for need. This is witnessed in cities characterized by rapid population turnover and redevelopment of buildings through urban regeneration. However, this key area of policy development is yet to be fully understood, for example, while preparing the material for this chapter, the authors identify with the paucity of research noted by Pani-Harreman et al. (2021) exploring the experience of ageing in place over time. Within their multiple case studies of older people living in neighbourhoods characterized by high levels of deprivation and rapid population change, they found that ageing in place is significantly influenced by an ever changing and dynamic life-course circumstances and the dynamics of families, communities, and neighbourhoods.

The local environmental context has become a significant factor in terms of the promotion of ageing in place, as reflected in the media campaigns by agencies promoting assisted living centres or older adults in India. However, there are clear and significant inequalities existing within the ageing population, (Rahman & Chittedi, 2022). Those who can make conscious decisions about where and with whom they live are dependent upon access to financial resources and support, the lack of which restricts decision making and the ability to maintain dignity and independence. Furthermore, ageing in neighbourhoods and communities that are perceived to be unsafe can generate exclusion, detachment, and a sense of 'being out of place' (Phillips et al., 2011).

The context or place of ageing has a profound effect on health, holistically referred to as the social determinants of health (SDH) (WHO, n.d). The SDH are described as those non-medical factors that influence health and reflect the social, economic, environmental, political, psychological, and spiritual influences that affect the person, family, and community. These include, for example, socioeconomic policies and systems, development agendas, social norms, social policies, cultural and political systems. Differences in health status can be seen and attributed to levels of income, health, and illness following a social gradient: the lower the socioeconomic position, the worse the health. As asserted by Marmot (2005) the degree of control an individual has over their life and the opportunities for social engagement and participation are crucial for well-being and longevity.

However, the Indian government as consequence of historically low public investment in health and social care and a substandard health and social care infrastructure is unable to fulfil the demands of the ageing population. The severely constrained public health system is unlikely to be able to respond to the challenges presented by the burden of communicable and noncommunicable diseases without a rethink of a public health strategy and significant investment. Nonetheless, public health policy must recognize the vulnerabilities and many of the challenges presented in old age are not a consequence of biological age alone but an interweaving system of inequalities.

The healthcare system in India is divided into two main sectors, the private and the public. The private sector comprises an amalgamation of hospitals and private clinics, usually in urban areas. According to the World Health Organization (2022), the poor health infrastructure and underfunding by government of health and social care programmes have been a major source of concern, both at the national and state levels. Consequently, the public Indian healthcare system is lacking in broad healthcare delivery, and because of that, the private sector has rapidly advanced its coverage among the Indian population further exacerbating inequalities.

This system of persistent inequality exacerbates the challenges faced by the over 60 age group who wish to remain within their own space. Palliative care is poorly developed, and secondly there are no unified organizations or institutions that provide care for those elderly people who live alone. This is largely due to the social structure and family culture that considers the parents should be cared for by their family members.

Within India, traditional family structure models are being challenged, due to the rising financial costs associated with daily living and the additional burden of reducing social security and increasing healthcare costs. The impact of this is illustrated by the increasing number of the elderly population left alone in rural areas reflecting the migration drift of younger family members to urban communities seeking greater opportunities and wellbeing. Agarwal et al. (2020) challenge the tenets of Indian culture and values. The combined effect of increased longevity and falling fertility has dramatically increased the size of the population group aged 60 years plus. This has contributed to unprecedented health, social, and economic challenges that are multifaceted and complex. These challenges are exacerbated by the substantive part of Indian society that live in poverty and remain unable to prepare for old age. This presents a significant challenge for policymakers who must respond proactively and as a matter of urgency to prevent an unprecedented ageing disaster.

Elderly people who have no family members are unlikely to receive state support. A key element of ensuring and maintaining independence is the provision of an appropriate support package and the provision of care to enable independence within the desired space. Within India the notion of the 'live in support carer' is very different to the European role. In India, a live in servant fulfils this role, but are not trained or qualified in health and social support. But they are not trained in palliative care and are not qualified. Their role is to carry out household duties and not health and social care related tasks.

Therefore, it is essential that a clear and accurate assessment of health and social needs is conducted, and the appropriate needs led plans put in place. Undoubtedly, attention must be given to the dynamic nature of place and its relationship with ageing people. Furthermore, a policy rethink needs to take place and to consider within a social model of health how ageing in place can be supported, embracing the needs of an ageing community and a changing society.

11.5 Conclusion

This chapter has examined the demographic changes that China and India have faced and scrutinized how both countries deal with ageing in a public policy context. To begin with, the authors focused on demographic comparisons, and as demonstrated, China's and India's demographics have evolved over time. In summary, given the demographics, it seems unlikely that China will regain its position as the most populated country as, in comparison with India, China's population is comparatively older and has a lower birth rate. Equally, while China may face issues caring for an ageing population, India will face issues related to managing population density, improving life expectancy, and increasing educational attainment.

Then, from a public policy perspective, the authors of this chapter examined how both countries adopt the ageing in place approach. In the context of China, in the past decade, China has developed rapidly in both policy and practice related to ageing in place, accumulating numerous successful experiences while also facing many challenges. Future policies and practices need to balance efficiency and fairness, focusing on improving the quality and efficiency of home-based care services while addressing shortcomings, promoting the sustainable development of home-based care services, treating the elderly as participants rather than clients (Halsall et al., 2023), and developing services of higher quality that are more efficient and more caring. In the context of India, the place of ageing undeniably has a significant effect on health and wellbeing; without generating an awareness of the vulnerabilities of this key and growing population group, persistent inequalities will be exacerbated. Policymakers must adopt a more holistic perspective by embracing a social determinant model approach to their planning and policymaking. However, without a policy rethink, adequate funding, and appreciation of the changing nature of family structure, values, and culture, the ageing population will continue to experience persistent challenge and inequalities.

Review Questions

1. How has economic transformation taken place in China and India?
2. Can you summarize the demographic changes in China and India?
3. Define what is meant by ageing in place.
4. What are the key characteristics of ageing in China and India?
5. Why is the ageing issue so important on public policy?

Acknowledgements The ideas for this chapter were originally presented at Huddersfield Business School's 8th Annual Research Conference at the University of Huddersfield, UK. The authors of this work have specific research interests in China's and India's socioeconomic status from a public policy perspective. The authors would like to express their thanks to Dr. Stefanie El Madawi who acted as our proof-reader.

References

Agarwal, A., Lubet, A., Mitgang, E., Mohanty, S., & Bloom, D. E. (2020). Population aging in India: Facts, issues, and options. In J. Poot & M. Roskruge (Eds.), *Population change and impacts in Asia and the Pacific* (pp. 289–311). Springer.

Barati, F., Khoddam, H., & Modanloo, M. (2022). Concept development of "Ageing in place": Application of hybrid model. *Journal of Family Medicine and Primary Care, 11*(11), 6646–6653.

Bardhan, P. (2010). The paradigm of capitalism under a developmental state: Does it fit China and India? *Singapore Economic Review, 55*(2), 243–251.

Bardhan, P. (2020) The Chinese governance system: Its strengths and weaknesses in a comparative development perspective. *China Economic Review*, 61, June.

Bigonnesse, C., & Chaudhury, H. (2020). The landscape of "aging in place" in gerontology literature: Emergence. *Theoretical Perspectives, and Influencing Factors, Journal of Aging and Environment, 34*(3), 233–251. https://doi.org/10.1080/02763893.2019.1638875

Bloom, D. E., Sekher, T. V., & Lee, J. (2021). Longitudinal Aging Study in India (LASI): New data resources for addressing aging in India. *Nature Aging, 1*, 1070–1072. https://doi.org/10.1038/s43587-021-00155-y

Budolfson, M., & Spears, D. (2021). Population ethics and the prospects for fertility policy as climate mitigation policy. *The Journal of Development Studies, 57*(9), 1499–1510. https://doi.org/10.1080/00220388.2021.1915481

Cheruvalath, R. (2015). Is the right of children to free and compulsory education act really beneficial to the poorer children in India? An analysis with special reference to the admission of poorer children in public unaided schools. *Education 3–13, 43*(6), 621–629. https://doi.org/10.1080/03004279.2013.847970

Cook, I. G., & Halsall, J. P. (2012). *Ageing in comparative perspective: Processes and policies*. Springer.

Cook, I. G., Halsall, J. P., & Wankhade, P. (2015). *Sociability, social capital, and community development: An international health perspective*. Springer.

Feng, W., Cai, Y., & Gu, B. (2013). Population, policy, and politics: How will history judge China's one-child policy? *Population and Development Review, 38*(s1), 115–129. https://doi.org/10.1111/j.1728-4457.2013.00555.x

Feng, W., Gu, B., & Cai, Y. (2016). The end of China's one-child policy. *Studies in Family Planning, 47*(1), 83–86. https://doi.org/10.1111/j.1728-4465.2016.00052.x

Fu, W., & Larbi, F. O. (2022). The rebirth of the chinese higher education: Analysis of higher education policies from the economic reform to date. *International Journal of Comparative Education and Development, 24*(1), 1–19. https://doi.org/10.1108/IJCED-04-2021-0044

Goel, A., & Kaur, A. (2022). The state of elderly in India: Life and challenges in the decade of healthy aging. *Indian Journal of Public Health, 66*(1), 1–2. https://doi.org/10.4103/ijph.ijph_190_22

Government of India. (2021). *Elderly in India in 2021*. Ministry of Statistics and Programme Implementation.

Halsall, J. P., Cook, I. G., Siddiqui, K., Kayama, D., Snowden, M., Stockton, J., & Holland, C. (2023). *Global public policy shifts: Comparative thoughts on ageing and social care*. Springer.

Haque, M. A., & Afrin, S. (2022). Active ageing index in Bangladesh: A comparative analysis with a European approach. *Ageing International, 47*, 447–464.

Huang S., & Sun S. (2021). The Development Process, Practical Difficulties and Path Optimization of Community-based Home Care in China. *China National Conditions and Strength, 10*, 9–13.

IMF. (2020). *Economic outlook update, June*. IMF. https://www.imf.org/en/Publications/WEO/Issues/2020/06/24/WEOUpdateJune2020. Accessed 20 May 2022.

Jha, P. K., & Rani, G. (2016; 2015). *Right to education in india: Resources, institutions and public policy*. Routledge. https://doi.org/10.4324/9781315666440.

Ji, J. (2022). Effect of home-based elderly care services on the quality of life for the elderly and policy implications in China. *Population and Development, 28*(03), 90–98.

Karlin, N. J., & Weil, J. (2017). Healthy ageing in a global context: Comparing six countries. *Ageing International, 42*, 1–22.

Lalchandani, K., Gupta, A., Srivastava, A., Usmanova, G., Maadam, A., & Sood, B. (2021). Role of financial incentives in family planning services in India: A qualitative study. *BMC Health Services Research, 21*(1), 1–905. https://doi.org/10.1186/s12913-021-06799-1

Lancet, T. (2022). Population ageing in China: Crisis or opportunity? *The Lancet (British Edition), 400*(10366), 1821–1821. https://doi.org/10.1016/S0140-6736(22)02410-2

Lu, J. (2022). Main types, practical support, and development trends of rural elderly care in China. *China Party and Government Officials Forum, 09*, 58–61. https://doi.org/10.14117/j.cnki.cn11-3331/d.2022.09.021

Macrotrends. (2023a). *India fertility rate 1950–2023*. https://www.macrotrends.net/countries/IND/india/fertility-rate. Accessed 10 Mar 2023.

Macrotrends. (2023b). *China fertility rate 1950–2023*. https://www.macrotrends.net/countries/CHN/china/fertility-rate. Accessed 10 Mar 2023.

Marmot, M. (2005). Social determinants of health inequalities. *The lancet, 365*(9464), 1099–1104.

Pani-Harreman, K., Bours, G., Zander, I., Kempen, G., & Van Duren, J. (2021). Definitions, key themes, and aspects of 'ageing in place': A scoping review. *Ageing & Society, 41*(9), 2026–2059.

Parkin, B., & Singh, J. (2022). India's uneven economic rebound creates winners and losers. *Financial Times*, March 7, London. https://www.ft.com/content/b99a4329-d79f-4512-9076-c76b0f50f550. Accessed 15 May 2022.

Phillips, J., Walford, N., & Hockey, A. (2011). How do unfamiliar environments convey meaning to older people? Urban dimensions of place lessness and attachment. *International Journal of Ageing and Later Life, 6*(2), 73–102.

Powell, J. (2023a). *New perspectives on health and social care*. Springer.

Powell, J. (2023b). *Sociology of ageing and death*. Springer.

Rahman, K. T., & Chittedi, K. R. (2022). Socioeconomic inequalities in healthy ageing in different regions of India: Evidence from the Longitudinal Ageing Study in India. *Review of Development and Change, 27*(2), 170–189.

Rapp, T., Ronchetti, J., & Sicsic, J. (2022). Where are populations Ageing better? A global comparison of healthy Ageing across organization for economic cooperation and development countries. *Value in Health, 25*(9), 1520–1527.

Reserve Bank of India (RBI). (2021). *Annual report*, May, Government of India: Reserve Bank of India. https://rbi.org.in/Scripts/AnnualReportMainDisplay.aspx. Accessed 12 May 2022.

Robinson, A. (2019). *India: A short history*. Thames & Hudson.

Siddiqui, K. (2015). Agrarian crisis and transformation in India. *Journal of Economics and Political Economy, 2*(1), 3–22.

Siddiqui, K. (2016). Will the growth of the BRICs cause a shift in the global balance of economic power in the 21st century? *International Journal of Political Economy, 45*(4), 315–338.

Siddiqui, K. (2018a). The political economy of India's post-planning economic reform: A critical review. *World Review of Political Economy, 9*(2), 235–264.

Siddiqui, K. (2018b). The political economy of India's economic changes since the last century. *Argumenta Oeconomica Cracoviensia, 19*, 103–132.

Siddiqui, K. (2020a). "A comparative political economy of China and India: A critical review" (Edi) Young-Chan Kim. In *China-India relations: Geo-political competition, and economic cooperation* (pp. 31–58). Springer.

Siddiqui, K. (2020b). The impact of Covid-19 on the global economy. *World Financial Review*, May–June, 25–31.

Siddiqui, K. (2021). Agriculture, sustainable development, and the government policy in the developing countries. *World Financial Review*, January–February, 44–59.

South, S. J., Trent, K., & Bose, S. (2014). Skewed sex ratios and criminal victimization in India. *Demography, 51*(3), 1019–1040. https://doi.org/10.1007/s13524-014-0289-6

Statista. (2023). *India: Life expectancy 1800–2020*. https://www.statista.com/statistics/1041383/life-expectancy-india-all-time/
Tang, D., Gao, X., Cai, J., & Coyte, P. C. (2022). The effect of China's two-child policy on the child sex ratio: Evidence from shanghai, China. *Inquiry (Chicago), 59*, 469580211067933–469580211067933. https://doi.org/10.1177/00469580211067933
The Economist. (2023). *The world ahead*. London.
The Financial Times. (2022). China's slowing growth will have a global impact, January 18, London.
United Nations. (2019). Department of Economic and Social Affairs, Population Division. *World population prospects 2019*, Online Edition, Rev. 1, https://go.nature.com/3cFqF0R
United Nations Department of Economic and Social Affairs. (2023). *India to overtake China as world's most populous country in April 2023, United Nations projects*. https://www.un.org/en/desa/india-overtake-china-world-most-populous-country-april-2023-united-nations-projects#:~:text=24%20April%202023%20%2D%20China%20will,the%20population%20of%20mainland%20China. Accessed 2 May 2023.
WHO. (2022). *Ageing and health*. https://www.who.int/news-room/fact-sheets/detail/ageing-and-health. Accessed 2 May 2023.
Wiles, J. L., Leibing, A., Guberman, N., Reeve, J., & Allen, R. E. (2012). The meaning of "ageing in place" to older people. *The Gerontologist, 52*(3), 357–366.
World Bank. (2023). *Fertility rate: India 1960–2020*. https://data.worldbank.org/indicator/SP.DYN.TFRT.IN?locations=IN. Accessed 2 May 2023.
World Data. (2023). *Average age by country*. https://www.who.int/india/health-topics/ageing. Accessed 2 May 2023.
World Health Organisation. (2023). *Ageing and health India*. https://www.who.int/india/health-topics/ageing. Accessed 2 May 2023.
World Population Review. (2023). *Countries by population density*. https://worldpopulationreview.com/country-rankings/countries-by-density. Accessed 2 May 2023.
World Population Review 1. (2023). *Total fertility rate*. https://worldpopulationreview.com/country-rankings/total-fertility-rate. Accessed 2 May 2023.
World Population Review 2. (2023). *World population by country*. https://worldpopulationreview.com/. Accessed 2 May 2023.
Wu, C. (2019). Research on "urban community home-based elderly care" in China. *Journal of Zhejiang Gongshang University, 03*, 91.
Yang, C. (2019). Research on the development of community home-based aged care policy in China: Analysis based on national policy texts from 1982 to 2018. *Journal of Anhui Administrative College, 02*, 104–112. https://doi.org/10.13454/j.issn.1674-8638.2019.02.016
Yao, Y., Hu, B., & Liu, X. (2023). Life expectancy trends in China in the post-COVID-19 era. *The Lancet Public Health*. https://doi.org/10.1016/S2468-2667(23)00074-9
Yi, Z., Vaupel, J. W., & Zhenglian, W. (1993). Marriage and fertility in China: 1950–1989. *Genus*, 17–34.
Zhang, P. (2016). Embedded elderly care: Research on innovation of elderly care service models in Shanghai. *Modern Management Science, 06*, 64–66.
Zhu, Q. (2017). Exploration and thinking on the development of embedded elderly care services in Shanghai communities. *China Civil Affairs, 16*, 30–32.

Chapter 12
'Sustainable Ageing' in a World of Crises

Nestor Asiamah

Acronyms

ATA	Activity theory of ageing
COVID-19	Coronavirus disease 2019
CTA	Continuity theory of ageing
DTA	Disengagement theory of ageing
EC	European Commission
MeSH	Medical Subject Headings
PA	Physical activity
SDGs	Sustainable Development Goals
SST	Socioemotional selectivity theory
UN	United Nations
WHO	World Health Organization

12.1 Introduction

Wellbeing for all is the core interest of stakeholders such as the World Health Organization (WHO) and governments, and this shared interest has brought about local, national, and international public health interventions. The WHO is renowned for championing several programmes aimed at the prevention of disease or the maintenance of health, examples being the WHO Physical Activity Strategy for the

N. Asiamah (✉)
Division of Interdisciplinary Research and Practice, School of Health and Social Care, University of Essex, Colchester, Essex, UK

Department of Gerontology and Geriatrics, Africa Centre for Epidemiology, Accra, Ghana
e-mail: n.asiamah@essex.ac.uk; nestor.asiamah@ace-gh.org

N. Asiamah et al. (eds.), *Sustainable Neighbourhoods for Ageing in Place*,
https://doi.org/10.1007/978-3-031-41594-4_12

European Region and the more recent 'Decade of Healthy Ageing (2020–2030)' initiative (Black & Oh, 2021; Rudnicka et al., 2020). These and similar programmes have been founded on the principle of collective action that emphasizes the involvement of all stakeholders in prioritized courses of action. To this end, the Decade of Healthy Ageing campaign was founded with key stakeholders including all member states of the United Nations (UN) (Rudnicka et al., 2020). While this approach to healthy ageing is promising, its momentum may be resisted or at least slowed by worsening global crises (i.e. climate change, infectious diseases, extreme industrialization, as well as violence and war).

War and violence, climate change events (e.g. flooding and extreme weather), unfavourable industrial practices resulting in land and air pollution, and the outbreak of infectious diseases such as the coronavirus disease 2019 (COVID-19) are an ongoing global issue. The current economic difficulty resulting from the outbreak of COVID-19 and the Russia-Ukraine war (Tosun & Eshraghi, 2022) is a resurgence of several previous episodes of economic recession, a typical example being the 2007–2009 global economic downturn (Jagannathan et al., 2013). These and similar experiences over the last century suggest that the world may plunge further into social, economic, and environmental crises. Yet, there are differences in how individuals, segments of the population (e.g. older adults), and countries are experiencing these problems.

Researchers (Asiamah, 2022; Asiamah et al., 2023; Chen et al., 2021; Kohrt et al., 2012) agree that some population groups such as children and older adults are more vulnerable to the foregoing crises. For instance, mortality and morbidity due to air pollution from radical industrial activities, extreme weather, and infectious diseases such as COVID-19 are more probable in older populations (Asiamah, 2022; Asiamah et al., 2023; Chen et al., 2021). These crises may affect humans for a long time; hence, current healthy ageing campaigns can be backdropped by them. Though these crises have been with humans for a long time, the enormity of their impact on society increases over time. This being so, the current approaches to healthy ageing must be resilient against them and must be framed to yield optimum outcomes in every context and dispensation. Healthy ageing campaigns may be the best way to respond to population ageing, but little is known about whether their underlying stakeholder efforts and the ageing process itself can be challenged by the crises.

In this chapter, therefore, the author proposes the concept of 'sustainable ageing' to guide multidisciplinary efforts that prevent the crises or minimize their impact on campaigns and the ageing process. Recommendations for stakeholder action are also provided. This chapter consists of seven sections. The first section, the introduction, provides a background to the chapter, whereas the second section sets the basis for defining the concept of sustainable ageing by providing a theoretical framework. The author then sets out the significance of the new concept in the third section and identifies its seven dimensions or thematic pathways in the fourth section. Implications for ageing in place are delineated in the fifth section, whereas limitations and future research directions are presented in the sixth section. The seventh section presents concluding remarks.

12.2 Theoretical and Conceptual Foundations for Sustainable Ageing

This section reviews relevant ageing initiatives, concepts, and theories that provide a basis for the concept of sustainable ageing.

12.2.1 Landmark Healthy Ageing Initiatives

Though the active ageing concept and its campaigns date back to the 1940s (Boudiny, 2013), the author focuses on the active ageing initiative championed by the WHO and European Commission (EC) in the late 1990s (Boudiny, 2013). This programme is of interest in this chapter as it is more recent and is still dictating policy directions globally. The initiative aims to provide a strategic framework guiding individuals and populations to maintain and engage with life and health over the life course (Black & Oh, 2021; Kalache & Gatti, 2003). To this end, individuals are expected to maintain engagement in the wider society and economy into later life, which is necessary for sustainable financial independence in older age. It promises to reduce the old-age dependency ratio, which is the number of older adults (e.g. people aged 65 and over) per 100 people of working age between ages 20–64 (Muszyńska & Rau, 2012). Reducing this ratio as much as possible lessens the economic burden on younger people and facilitates economic development.

What makes the active ageing initiative remarkable is its influence on other health promotion programmes such as the WHO PA Strategy for the European Region (OMS, 2015; World Health Organisation, 2021), which provides a framework of strategies and recommendations for enabling individuals to maintain an active lifestyle in the interest of active ageing. Noteworthy within this framework is an evidence-based collection of PA recommendations formulated for different age groups. A recommendation for older adults, for example, is 75–150 minutes of vigorous-intensity PA or 150–300 minutes of moderate-intensity PA per week (Okely et al., 2021). Meeting this or a relevant recommendation is necessary for avoiding physical inactivity and maintaining wellbeing into older age (Okely et al., 2021; World Health Organisation, 2021).

Overlapping (in purpose) with active ageing and PA promotion campaigns is the Decade of Healthy Ageing (2016–2020) initiative by the WHO, its member states, and stakeholders of the Sustainable Development Goals (SDGs) (Rudnicka et al., 2020). The programme had five objectives, the first of which was a commitment to actions intended to allow everyone to age in good health in every country. This objective was informed by a recognition of inequity and diversity in the way individuals and populations age in different contexts. For instance, some older adults in their 80s in a country can be as healthy as younger adults in their 30s in this country, which signifies diversity in ageing. Older adults in their 70s may show significantly different health profiles, which signifies inequity. Given that inequity and diversity were not eliminated by the end of 2020, the programme was renewed.

The Decade of Healthy Ageing (2020–2030) is the renewed version of the above strategy and accompanied ten objectives, including aligning health systems to the needs of older adults (Dixon, 2021; Rudnicka et al., 2020). The tenth objective is to extend the global network for age-friendly cities and communities, which advances the WHO's network of age-friendly cities and communities launched in 2010 (Dixon, 2021; Keating, 2022; Rudnicka et al., 2020). This network is geared toward the provision of safe, inclusive, and sustainable cities and human settlements, especially for older adults. Now covering more than 1000 cities and communities worldwide (Rudnicka et al., 2020), the network prioritizes creating access to green public space and improving social and built environment factors (e.g. street connectivity, mixed land use, high residential density, parks, traffic signs, disability-friendliness of the built environment) (Black & Oh, 2021; Lowen et al., 2015) that support social participation among older adults. By the above tenth objective, the WHO and its partners demonstrate an unwavering commitment to the provision of age-friendly contexts. Worth noting in the next section are key concepts that have become prominent in the gerontological literature owing to the foregoing initiatives.

12.2.2 Flagship Concepts in Gerontology

Active ageing is a landmark concept in gerontology and helps to understand other concepts such as healthy ageing and successful ageing. Active ageing means individuals staying in charge of their lives for as long as possible and maintaining their contribution to society and the economy if possible (Dixon, 2021; Rudnicka et al., 2020). It is also defined as the process of optimizing opportunities for health, social participation, and security to enhance the quality of life and wellbeing as people age (Boudiny, 2013). As these definitions suggest, the maintenance of physical functional ability necessary for active participation in the society and economy in later life is central to active ageing.

Healthy ageing, on the other hand, is 'the process of developing and maintaining the functional ability that enables wellbeing in older age' (Peel et al., 2004). Thus, healthy ageing requires the maintenance of functional abilities necessary for health later in life. Judging by the foregoing definitions, an attribute shared by active and healthy ageing is the maintenance of functional ability, which is necessary for engaging in the economy and society in later life. Moreover, both concepts have been closely associated with 'ageing well', which refers to the maintenance of wellbeing or health over the life course (Boudiny, 2013). Yet, healthy ageing is interpreted to place more emphasis on functional ability and its maintenance over time to delay or avoid the onset of disease (Foster & Walker, 2021). This idea recalls another related concept, successful ageing, which means high physical, psychological, and social functional capacity in later life or old age without major diseases (Bowling & Dieppe, 2005).

Like active and healthy ageing, successful ageing involves illness avoidance, engagement with life, and maintaining functional ability in later life.

Another unique concept typical of environmental gerontology is 'ageing in place' or 'ageing in context' (Asiamah et al., 2023), which is about the maintenance of health and functional ability in one's home and neighbourhood and avoiding long-term admission into a care or nursing facility due to a disease or physiological limitation (Asiamah et al., 2023; Tao et al., 2021). Some argue that ageing in place includes keeping a sense of attachment to one's neighbourhood or home (Peace et al., 2011). While ageing in place has been framed differently by different authors, researchers agree that it includes or requires the maintenance of optimal health and functional ability (Asiamah et al., 2023; Tao et al., 2021), which makes it possible for people to avoid admission into long-term institutional care or growing old in unfamiliar settings they are unhappy with. 'Positive ageing' and 'productive ageing' (Boudiny, 2013) have also been used in the literature to refer to the maintenance of wellbeing, engagement with life, and functional capacity, which means that all the above types of ageing include the maintenance of engagement with life, functional ability, and wellbeing, over the life course.

12.2.3 The Theoretical Foundation

The theoretical basis of the active ageing initiative is the activity theory of ageing (ATA) (Havighurst, 1961), which asserts that individuals can maintain PA, physical functional ability, and health in the ageing process by adapting previous life experiences and resources such as social networks. Being able to adapt past experiences includes resilience (Cosco et al., 2017; Stern et al., 2019), which people need as they age to overcome barriers to social and physical activities. The ATA adds that ageing people can maintain independence by maintaining control over all aspects of their lives, including paid or unpaid employment. This potential to exercise control over one's life and maintain participation in society and the economy is said to be implicit in autonomy (Heathcote, 2000; Van Der Cammen et al., 2017). Deductively, the original campaign of active ageing championed by the EC and WHO values the maintenance of physical functional ability, health, and autonomy through adaptive behaviour facilitated with resilience. The argument of ATA is a parallel of the import of the Continuity Theory of Ageing (CTA) developed by Robert Atchley, which advocates the continuity of a normal life characterized by health-seeking behaviours (e.g. healthcare utilization, social participation, and PA), health, and functional capacity (Abramowska-Kmon et al., 2023; Atchley, 1989). Even so, both the CTA and ATA are at odds with the disengagement theory of ageing (DTA).

The DTA is arguably the most popular framework directly opposed by the ATA and CTA (Asiamah, 2017; Menassa et al., 2023), though it was published before the ATA. It argues that the individual's PA and health dwindle over the life course due to a gradual decline in physical functional ability and other personal resources such as social networks and income. Though this theory has been criticized for its negativity and for undermining adaptability, resilience, and autonomy as applied to the ATA, researchers have endorsed its plausibility and validity in some contexts, including settings with high income inequalities, limited access to health services, and low walkability (Asiamah, 2017).

Another theory that holds positive and negative views about older adults and ageing is the socioemotional selectivity theory (SST) (Cleveland & Agbeke, 2019; Löckenhoff & Carstensen, 2004). The negative view of this theory is how it limits social and physical activities in older populations to a few tasks tied to older adults' future time perspective, which concerns people's perceptions of the brevity of their remaining life (Bardach et al., 2021; Löckenhoff & Carstensen, 2004). The theory explains that older adults are aware of how short and important their remaining life is and, therefore, make prioritized choices that may limit their social and physical environment. For instance, older adults, impelled by their future time perspective and a change in life goals (Asiamah et al., 2021a), decide to spend time with only loved ones such as in-laws and grandchildren and avoid the wider society.

The SST's positive view of ageing can be used to explain the utilization or diffusion of health-supporting information technology ability among older adults. People's future perspective instils in them a sense of urgency and a need to prioritize social and economic activities that are exclusively rewarding in later life. For example, though older adults are usually unwilling to use information technologies in general, they may be prepared to use those that support their life or improve their longevity (Khosravi et al., 2016; Liu et al., 2016; Sghaier et al., 2022; Yusif et al., 2016). Technologies enabling older adults to receive healthcare (e.g. telemedicine) and perform activities of daily living and instrumental activities of daily living can be particularly valued by them. Moreover, human development programmes can be effective at increasing older adults' information technology adoption and ability (Klaming et al., 2023; Liu et al., 2016; Yusif et al., 2016). No doubt, this aspect of the SST can guide efforts to enable older adults to use information technologies, which are a key part of the Decade of Healthy Ageing (2020–2030) strategy.

Though at odds with the DTA, the ATA and CTA are seen as valid frameworks for taking optimistic views of the ageing process and can, therefore, be considered valid theoretical compasses guiding ageing initiatives such as the active ageing campaign of the late 1990s. For this reason, their argument about the ability of individuals to maintain PA, health, functional ability, and autonomy through adaptive behaviour is adopted in this chapter to operationalize the concept of sustainable ageing. The next section discusses why sustainable ageing as a concept must be framed in contemporary gerontology.

12.3 Sustainable Ageing: Why It Matters in Contemporary Gerontology

The discussion so far has shown an emphasis on the central role of wellbeing, engagement with life, and physical functional ability in ageing; a need for people to maintain these three conditions is evident in the previous section. Though the roles of resilience, autonomy, and similar attributes in ageing are recognized by these concepts and theories (Cosco et al., 2017; Heathcote, 2000; Menassa et al., 2023), these depend on functional ability and wellbeing. Thus, maintaining wellbeing and functional ability is the hallmark of healthy, successful, and active ageing as well as ageing in place. Even so, the global crises and their worsening economic difficulties may make it difficult for people to maintain wellbeing as they age and may render ageing interventions less effective.

A growing body of literature recognizes the devastating impacts of war and violence on communities around the world. Some studies (Oleksiyenko et al., 2021; Quinn et al., 2022; Strong et al., 2015) have reported an increase in public health risks due to war and violence in various countries, and casualties from wars and violence include new cases of physical disability (Baker, 2014; Polack et al., 2021). Studies agree that older adults and children are among the most affected and vulnerable (Benevolenza & DeRigne, 2019; Malik et al., 2018). Many infectious diseases such as cholera have claimed many lives globally (Lopez et al., 2015), and the awful memories of the socioeconomic and public health emergency caused by the COVID-19 pandemic in the past few years are far-reaching. Older adults also experienced increased social isolation and mental health problems during the COVID-19 pandemic (Berg-Weger & Morley, 2020; Ganz et al., 2020).

The adverse impacts of extreme industrial activities on wellbeing have been recently reported (Asiamah, 2022; Li et al., 2018; Ntengwe, 2006). Air and noise pollution from industrial practices within cities and communities can destabilize community life (Asiamah, 2022; Ntengwe, 2006), especially among older adults with underlying health conditions such as asthma (Asiamah, 2022). The gravity of these and other negative impacts can be expected to be higher in contexts experiencing multiple crises. Commentators (Asiamah et al., 2021b; Borio, 2020; Gandasari & Dwidienawati, 2020) have alluded that the global crises would not only stay with humanity for a long time but may also trigger major episodes of the macroeconomic downturn. Since older adults live on the least income (Ku & Kim, 2020; Olivera & Tournier, 2016), they may be the most hard-hit population by the economic difficulties emanating from the crises. Similarly, economic difficulties complemented by the crises can delay or halt active investments in healthy ageing programmes.

The main problem with the global crises is their potential to degrade the environment where people age (Asiamah, 2022; Li et al., 2018), undermine the normalcy of life for older adults (Asiamah, 2022), and weaken collaboration as well as public health expenditure and spending (Qin et al., 2022). Undoubtedly, the crises and their associated economic difficulties are a new and growing dimension of the hurdle of attaining equity in health and healthy ageing. This dimension alone is big enough to overwhelm, frustrate, and thwart efforts toward realizing the goals of the Decade of Healthy Ageing (2020–2030) and even the SDG for health (i.e. SDG-3). This being so, it is necessary to ponder sustainable ageing, a potential new way to support people to age in health and in place. The next section presents the authors' framing of the sustainable ageing.

12.4 Sustainable Ageing as a Potential Way Forward

The author proposes sustainable ageing as a new approach to enabling people to age well with no or little impact of the global crises. At the individual level, sustainable ageing is the maintenance of a high physical functional ability and engagement with life necessary for wellbeing in later life with no or little impact of the crises. This definition suggests that people's sustainable ageing results in a high functional capacity, engagement with life, and health in older age, though they may live in communities experiencing the crises. Engagement with life refers to active participation in society (e.g. social participation) and the economy (e.g. paid employment or volunteering) in later life (Liffiton et al., 2012). Even if people decided to retire from work while in good health, they had a high financial potential for meeting needs over their remaining lifespan. The phrase 'no or limited impact of the crises' suggests two ideas. Firstly, ageing people can avoid the crises through personal efforts (e.g. taking precautions not to be harmed in a civil conflict) or preventive mechanisms instituted by stakeholders such as early detection of a tsunami and supporting potential victims to evacuate. Secondly, these or similar efforts can minimize the impact of the crises. For example, people can learn about what to do during natural disasters and apply what they have learned to avoid serious injuries or death.

Sustainable ageing can be optimized through a 'co-design strategy', which the author views as a multidisciplinary effort to enable individuals and populations to age well. This effort is driven by the 'principle of co-design' that emphasizes the involvement of all stakeholders (e.g. individuals, governments, opinion leaders, and civil society organizations) in healthy ageing programmes. Individuals can choose the best ways to age such as maintaining a life-long trajectory of health-seeking behaviours through the utilization of public systems (e.g. services and parks). Governments can collaborate with their public health stakeholders and agencies (e.g. WHO and Ministries of Health) to provide a public health infrastructure (e.g. clinical and community care facilities, improving age-friendly neighbourhoods, and improving policies and their implementation) that is resilient against the crises. A key facet of the co-design strategy is a need for governments and other central stakeholders (e.g. WHO) to conscientize all stakeholders possibly through health education to play their respective roles. This effort would engender enough synergy against the crises and for sustaining campaigns for healthy ageing in contexts experiencing multiple crises and where individuals are needed in policy formulation and execution. Since the global crises can have a long-term effect on society (Asiamah et al., 2021b), sustainability of the co-design strategy is necessary, without which the momentum for success may collapse prematurely.

Table 12.1 shows the author's recommended actions for sustainable ageing and the thematic pathways to which they belong. While these actions are not exhaustive, they are the primary actions recognized in the literature for enabling people to age well with little or no impact of the crises. For example, the elimination of ageism is one of the ten objectives of the Decade of Healthy Ageing (2020–2030) strategy. Some experts (Krishnaswami et al., 2020; Pruchno, 2019; Rodeschini, 2011) also argue that the future of healthy ageing can only be bright if people are enabled to use information technologies effectively to achieve important goals. As mentioned earlier in this chapter, information technologies can help ageing people to perform key social, physical, and economic activities (Robbins et al., 2018; Yusif et al., 2016), which signifies their important role in sustainable ageing as shown in Table 12.1.

The first thematic pathway in Table 12.1 (i.e. a preventive approach to ageing) draws on studies proposing different approaches to the prevention of disease or the development of multiple morbidities in older age (Grande et al., 2020; Lunenfeld & Stratton, 2013). Preventive ageing is about efforts and programmes enabling people to avoid age-related morbidity and disability throughout their lifespan; people should not be supported only when they are already with a disease or disability in older age. It can reduce the cost of healthcare and its expenditure as well as economic losses due to the burden of disease among older adults.

'Partnerships against the crises and barriers' in the table is the author's emphasis on the principle of co-design that requires a multidisciplinary approach to sustainable ageing; interventions must be developed and executed through the collaborative efforts of politicians, civil society representatives, traditional or community leaders, and experts from relevant disciplines (e.g. engineers, surveyors, gerontologists, social workers, economists, human resources personnel, and health workers). An important part of the seventh pathway is a partnership for a financial base resilient against the crises for facilitating ageing programmes; this partnership may include public health organizations, governments, financial institutions, and philanthropic organizations. Sustained strength in financing ageing programmes is needed since the resource base of these programmes may be threatened by the crises.

Figure 12.1 is a framework depicting outcomes including sustainable ageing of the effective implementation of the thematic pathways. Physical functional ability and healthy behaviours are both an outcome of the thematic pathways and potential mediators of the relationship between the pathways and sustainable ageing. It is assumed that people would have optimal wellbeing in older age if they maintain healthy behaviours and functional capacity. Since engagement with life later in life requires physical functioning (Abud et al., 2022; Boudiny, 2013), it is a part of sustainable ageing in older age. Though functional ability can have a causal effect on sustainable ageing, ideally through experimental research designs, the framework depicts its correlation with sustainable ageing for two reasons. Firstly, wellbeing and engagement with life, especially earlier in life, are necessary for exercising functional capacity; people who fail to engage with life earlier in life are likely to experience a gradual decline in their wellbeing, which would result in low functional ability in older age. Finally, functional ability in later life can facilitate engagement with life and, thus, foster continuity in wellbeing.

The fourth part of the framework consists of personal factors (i.e. indicators of inequity and diversity) that can influence sustainable ageing. Age, for instance, can be associated with inequity in successful ageing in the sense that individuals in the same age group may report different levels of sustainable ageing possibly due to differences in socioeconomic factors (e.g. income, education, cultural orientation,

Table 12.1 The author's recommended thematic pathways and their key actions toward sustainable ageing in a world of crises

Description	Recommended action
A preventive ageing approach (Grande et al., 2020; Lunenfeld & Stratton, 2013)[a]	
Enabling ageing people[b] to maintain optimal health, functional ability, and financial independence into later life instead of waiting to support them later in life when they are with illnesses and physiological limitations	1. Maximize access to preventive healthcare for all
	2. Creating awareness about the need to consult a physician regularly at the individual level to avoid illness
	3. Allowing everyone to meet recommended PA levels
	4. Eliminating inequities in employment and income, especially in communities facing the global crises
	5. Introducing schemes to meet the financial needs of ageing people across the lifespan
Elimination of inequity and diversity (Carey & Crammond, 2015; Marmot et al., 2012; Rudnicka et al., 2020)[a]	
Helping people to age in good health regardless of their environment, race, culture, and socioeconomic status. This includes enabling them to age in the same positive ways in all contexts	1. Removing inequalities in income, employment, and health in relevant segments of the population
	2. Increasing funding for research investigating the impact of intersectionality on health
	3. Monitoring how different groups are ageing in their social and built environment
Equitable access to age-friendly cities and communities (Fitzgerald & Caro, 2014; Lowen et al., 2015; Rudnicka et al., 2020)[a]	
The provision and enhancement of age-friendliness of cities and communities and enabling all ageing people to maintain access to age-friendly neighbourhoods	1. Increasing the number of communities and cities involved in the WHO's age-friendly network
	2. Educating ageing people to use the built environment to maintain health-seeking behaviours such as PA and healthcare utilization
	3. Making neighbourhoods walkable in both rural and urban settings
	4. Reducing the environmental impact of the global crises
	5. Discouraging ageism and elderly abuse to maximize the inclusiveness of neighbourhoods

(continued)

Table 12.1 (continued)

Description	Recommended action
Drawing on technologies for ageing (Krishnaswami et al., 2020; Merkel & Kucharski, 2019; Pruchno, 2019)[a]	
Interventions that enable ageing people to use relevant information technologies to facilitate their health-seeking behaviours (e.g. PA and healthcare utilization). It includes improving peoples' willingness and ability to use information technologies to pursue personal interests and any health-seeking goal	1. Training ageing people to use information technologies
	2. Developing age-friendly information technologies
	3. Introducing technologies that support ageing people during the crises
	4. Enabling ageing people to avoid threats and security issues with information technologies
	5. Involving ageing people in the design of health-oriented technologies
Human development and health education (Chang et al., 2023; de Sam Lazaro et al., 2023; Klaming et al., 2023)[a]	
Developing and empowering ageing people to use information technologies as well as environmental and social resources (e.g. social networks) to avoid or minimize the impact of the global crises and other challenges to healthy ageing. This includes research-based interventions and training enabling people to overcome barriers to healthy ageing	1. Intensification of health education intended to enable ageing people to maintain health amidst the crises
	2. Promoting health-seeking behaviours and how to maintain them during the crises
	3. Improving ageing people's resilience and ability to adapt amidst crises through training
	4. Health education programmes intended to enhance ageing people's ability to maintain social, functional, and financial independence
	5. Improving retirement benefits and removing inequalities in them
Overcoming the global crises (success against the crises) (Barchielli et al., 2022; Stuchul et al., 2021)[a]	
Stakeholder interventions and individual efforts to avoid or minimize the impacts of the global crises on health as well as personal, social, and environmental factors (e.g. social networks and employment) that positively influence ageing	1. Recognizing the crises as a new dimension to the hurdle against healthy ageing efforts
	2. Developing global, national, and regional strategies to support people to age with or without the crises
	3. Integrating efforts against the crises into current healthy ageing campaigns
	4. Preventing the occurrence of crises where possible

(continued)

Table 12.1 (continued)

Description	Recommended action
Partnership against the crises and barriers (Barbabella et al., 2020; Khojasteh et al., 2022; Lagiewka, 2012)[a]	
Local, national, and international partnerships (including partnerships for sustainable financing of campaigns) for formulating and implementing policies toward healthy ageing. These partnerships partly concern the above thematic areas	1. Involving all stakeholders (e.g. governments, civil society, and ageing people) in developing healthy ageing policies against the crises
	2. Developing and securing a financial base resilient against the crises to facilitate ageing campaigns and interventions
	3. Involving ageing people and other vulnerable groups in the design of healthy ageing policies against the crises
	4. Localizing relevant partnerships to meet the unique needs of every community and country
	5. Constituting a central coordinating body of all relevant stakeholders, locally and internationally
	6. Routinely assessing the impact of collaborative efforts

[a]Key themes from the literature on which actions can be taken; [b]the term 'ageing people' refers to the general population but emphasizes older adults; preventive ageing supports all segments of the population, not only older adults, to age in good health; recommended actions are not listed in order of importance; *PA* - physical activity

and employment). The presence of gender in the framework suggests that men and women would report different levels of sustainable ageing due to possible differences in employment and income between them (Drydakis, 2022). Admittedly, the personal factors are not exhaustive, and other variables that may be associated with them and sustainable ageing can be infused into the model in future research. Both the model and suggested thematic pathways have implications for ageing in place as delineated as follows.

12.5 Implications for Ageing in Place

As earlier discussions suggest, ageing in place is about the maintenance of wellbeing and functional capacity as well as a sense of attachment to one's neighbourhood, which are necessary for spending the rest of life at home. It depends on how well the neighbourhood is designed to support health-seeking behaviours. The good thing is that current healthy ageing campaigns (e.g. a Decade of Healthy Ageing as well as WHO age-friendly cities and communities network) aim to provide age-friendly cities and communities (Rudnicka et al., 2020). This state-of-the-art coupled with the third pathway (i.e. equity in access to age-friendly cities and communities) signifies that sustainable ageing supports or includes ageing in place. Even so, ageing in place is more possible with the principle of co-design and the prioritization of investment in greener communities to preserve the natural features (e.g. wildlife, rivers, and green space) of neighbourhoods. This approach not only better supports active ageing but also aligns with the SDGs.

People cannot age in place if individuals are not involved in the co-design strategy for sustainable ageing. Ageing people can participate in this approach in a couple of ways. Firstly, stakeholders taking the lead (e.g. governments and WHO) ought to include residents in planning communities and their improvement; this is a client-focused approach that provides the relevant socio-environmental infrastructure through relationship-building processes involving end users of public infrastructure. Age-friendliness of communities cannot influence healthy behaviours and ageing in place if it undermines the interests and needs of residents. Including traditional and opinion leaders as well as local governments in planning and designing neighbourhoods can be an effective way to meet the needs of residents.

Worth acknowledging are recent calls (Danquah et al., 2022; Rodrigues & Franco, 2021) for residents to be responsible for the age-friendliness of their neighbourhoods. These calls emphasize a need for residents to pursue pro-environment behaviours (e.g. keeping the neighbourhood clean, regular maintenance of parks and gardens, and walking or bicycling instead of driving) that either add to or preserve the existing age-friendly infrastructure. Without this commitment from residents, age-friendly cities would deteriorate, especially in contexts where governments are unable to sustain their investment in age-friendly communities possibly due to the crises. This and other thoughts shared in this chapter, nevertheless, have some limitations as discussed in the next section.

12.6 Limitations and Future Research

The thematic pathways suggested in this chapter (see Table 12.1) are not exhaustive; only the major strategies identified in the literature are reported. The author, therefore, calls for future research aimed at developing a more holistic list of thematic areas and actions. Sustainable ageing is not proposed in this chapter as a replacement for the traditional concepts (e.g. healthy ageing, active ageing, successful ageing, and ageing in place) in gerontology; it only draws on the key ideas behind these concepts (e.g. the maintenance of physical functional ability, engagement with life, and health in later life) to propose a framework of actions for avoiding or

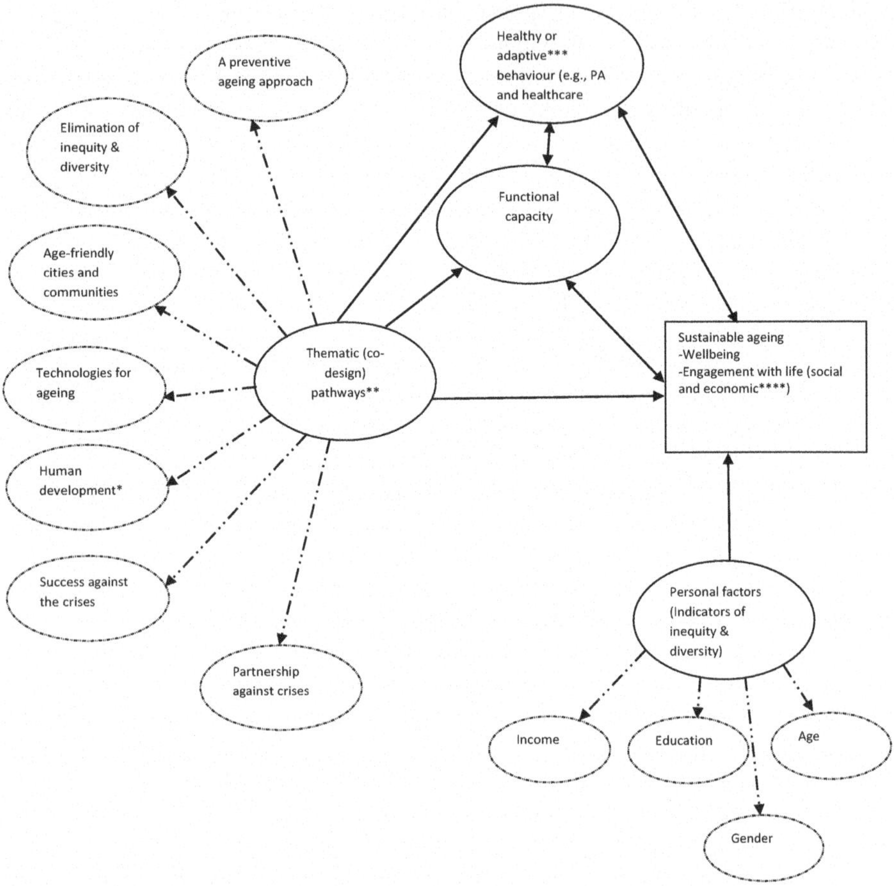

Fig. 12.1 The author's conceptual framework of the link between the thematic pathways and sustainable ageing.
****Engagement with life in later life ideally requires or includes social, financial, and functional independence;
***adaptive behaviour ideally requires or includes resilience;
**the thematic pathways can be correlated and ideally function as a unit but these potential correlations are not shown – success and partnership against the crises, for example, are necessary for sustained human development and other pathways;
*human development includes health education;
broken arrows represent dimensions; PA – physical activity

minimizing the adverse impacts of the crises on collaborative approaches to ageing and the ageing process itself. Finally, this chapter does not report considerations for assessing and measuring sustainable ageing and its relationship with other existing concepts. Yet, it provides key thoughts that researchers and stakeholders may consider in, respectively, framing research and making decisions.

12.7 Conclusions

Initiatives such as the Decade of Healthy Ageing are promising approaches to ageing, but these are being executed during a period of worsening crises. Coupled with economic difficulties, these crises are expected to thwart or weaken collaborative interventions enabling people to age well. Individuals are also expected to experience a more difficult ageing process that would require higher resilience and adaptive behaviour. Thus, the crises and their associated economic problems constitute a new dimension to the barriers to healthy ageing campaigns and the ageing process. Hence, approaches to sustainable ageing that prevent the crises or minimize their negative impact on collaborative campaigns and the ageing process are necessary. These approaches include a co-design strategy taking the form of multidisciplinary partnerships against the crises. A preventive pathway to ageing, which is one of these approaches, encompasses programmes enabling people to avoid age-related morbidity and disability throughout their lifespan. Sustainable ageing is the maintenance of functional ability, engagement with life, and wellbeing into later life with no or little impact of the crises.

Review Questions

1. Explain the concept of sustainable ageing, and discuss its unique significance vis-à-vis other concepts such as healthy ageing and active ageing.
2. Discuss the principle of co-design and its role in achieving sustainable ageing.
3. Enumerate and define the seven thematic pathways to sustainable ageing. Why may a preventive approach to ageing optimize cost at the national level?
4. Draw a diagram to illustrate the potential relationship between the thematic pathways, sustainable ageing, healthy behaviours, functional ability, and personal factors.
5. Discuss two ways you can play a role in preserving or improving age-friendly attributes in your neighbourhood.

References

Abramowska-Kmon, A., Mynarska, M., & Timoszuk, S. (2023). Coping strategies for happy childless ageing: An explorative study in Poland. *Genus, 79*(1). https://doi.org/10.1186/s41118-022-00180-8

Abud, T., Kounidas, G., Martin, K. R., Werth, M., Cooper, K., & Myint, P. K. (2022). Determinants of healthy ageing: A systematic review of contemporary literature. In *Aging clinical and experimental research* (Vol. 34, Issue 6, pp. 1215–1223). Springer Science and Business Media Deutschland GmbH. https://doi.org/10.1007/s40520-021-02049-w.

Asiamah, N. (2017). Social engagement and physical activity: Commentary on why the activity and disengagement theories of ageing may both be valid. *Cogent Medicine, 4*(1), 1289664. https://doi.org/10.1080/2331205x.2017.1289664

Asiamah, N. (2022). Walkable urban neighborhoods: The adverse effects of industrialization and climate change in developing countries. *Sustainable Urbanism in Developing Countries, March*, 247–262. https://doi.org/10.1201/9781003131922-16

Asiamah, N., Conduah, A. K., & Eduafo, R. (2021a). Social network moderators of the association between Ghanaian older adults' neighbourhood walkability and social activity. *Health Promotion International*, 1–11. https://doi.org/10.1093/heapro/daaa156

Asiamah, N., Opuni, F. F., Mends-Brew, E., Mensah, S. W., Mensah, H. K., & Quansah, F. (2021b). Short-term changes in behaviors resulting from COVID-19-related social isolation and their influences on mental health in Ghana. *Community Mental Health Journal, 57*(1), 79–92. https://doi.org/10.1007/s10597-020-00722-4

Asiamah, N., Bateman, A., Hjorth, P., Khan, H. T. A., & Danquah, E. (2023). Socially active neighborhoods: construct operationalization for aging in place, health promotion and psychometric testing. *Health Promot Int, 38*, 1–10.

Atchley, R. C. (1989). A continuity theory of normal aging. *Gerontologist, 29*(2), 183–190. https://doi.org/10.1093/geront/29.2.183

Baker, M. S. (2014). Casualties of the global war on terror and their future impact on health care and society: A looming public health crisis. *Military Medicine, 179*(4), 348–355. https://doi.org/10.7205/MILMED-D-13-00471

Barbabella, F., Cela, E., Di Matteo, C., Socci, M., Lamura, G., Checcucci, P., & Principi, A. (2020). New multilevel partnerships and policy perspectives on active ageing in Italy: A national plan of action. *International Journal of Environmental Research and Public Health, 17*(24), 1–17. https://doi.org/10.3390/ijerph17249585

Barchielli, B., Cricenti, C., Gallè, F., Sabella, E. A., Liguori, F., Da Molin, G., Liguori, G., Orsi, G. B., Giannini, A. M., Ferracuti, S., & Napoli, C. (2022). Climate changes, natural resources depletion, COVID-19 pandemic, and Russian-Ukrainian war: What is the impact on habits change and mental health? *International Journal of Environmental Research and Public Health, 19*(19). https://doi.org/10.3390/ijerph191911929

Bardach, S. H., Rhodus, E. K., Parsons, K., & Gibson, A. K. (2021). Older adults' adaptations to the call for social distancing and use of technology: Insights from socioemotional selectivity theory and lived experiences. *Journal of Applied Gerontology, 40*(8), 814–817. https://doi.org/10.1177/0733464821996864

Benevolenza, M. A., & DeRigne, L. A. (2019). The impact of climate change and natural disasters on vulnerable populations: A systematic review of literature. *Journal of Human Behavior in the Social Environment, 29*(2), 266–281. https://doi.org/10.1080/10911359.2018.1527739

Berg-Weger, M., & Morley, J. E. (2020). Loneliness and social isolation in older adults during the COVID-19 pandemic: implications for gerontological social work. *Journal of Nutrition, Health and Aging, 24*(5), 456–458. https://doi.org/10.1007/s12603-020-1366-8

Black, K., & Oh, P. (2021). Assessing age-friendly community progress: What have we learned? *The Gerontologist, 62*(1), 6–17. https://doi.org/10.1093/geront/gnab051

Borio, C. (2020). The Covid-19 economic crisis: Dangerously unique. *Business Economics, 55*(4), 181–190. https://doi.org/10.1057/s11369-020-00184-2

Boudiny, K. (2013). "Active ageing": From empty rhetoric to effective policy tool. *Ageing and Society, 33*(6), 1077–1098. https://doi.org/10.1017/S0144686X1200030X

Bowling, A., & Dieppe, P. (2005). What is successful ageing and who should define it? *British Medical Journal, 331*(7531), 1548–1551. https://doi.org/10.1136/bmj.331.7531.1548

Carey, G., & Crammond, B. (2015). Action on the social determinants of health: Views from inside the policy process. *Social Science and Medicine, 128*, 134–141. https://doi.org/10.1016/j.socscimed.2015.01.024

Chang, S. H., Chiang, C. C., & Chien, N. H. (2023). Efficacy of a multicomponent exercise training program intervention in community-dwelling older adults during the COVID-19 pandemic: A cluster randomized controlled trial. *Geriatric Nursing, 49*, 148–156. https://doi.org/10.1016/j.gerinurse.2022.11.019

Chen, Y., Klein, S. L., Garibaldi, B. T., Li, H., Wu, C., Osevala, N. M., Li, T., Margolick, J. B., Pawelec, G., & Leng, S. X. (2021). Aging in COVID-19: Vulnerability, immunity and intervention. *Ageing Research Reviews, 65*(August 2020), 101205. https://doi.org/10.1016/j.arr.2020.101205

Cleveland, J. N., & Agbeke, D. V. (2019). Socioemotional selectivity theory lifespan perspectives on job performance, performance appraisal/management and creative performance. 2009.

Cosco, T. D., Howse, K., & Brayne, C. (2017). Healthy ageing, resilience and wellbeing. *Epidemiology and Psychiatric Sciences, 26*(6), 579–583. https://doi.org/10.1017/S2045796017000324

Danquah, E., Asiamah, N., Opuni, F. F., Ocloo, E. C., & Ricky-Okine, C. K. (2022). Pro-environment behavioural moderators of the association between perceived walkability and social activity. *Journal of Transport and Health, 27*(June), 101533. https://doi.org/10.1016/j.jth.2022.101533

de Sam Lazaro, S. L., Nitschke Durben, A. M., & Kline, J. J. (2023). Barriers and opportunities to support the oral health of older adults: A rapid review of health policy and systems. *International Journal of Aging and Human Development, 96*(1), 51–62. https://doi.org/10.1177/00914150221106098

Dixon, A. (2021). The United Nations Decade of Healthy Ageing requires concerted global action. *Nature Aging, 1*(1), 2. https://doi.org/10.1038/s43587-020-00011-5

Drydakis, N. (2022). Sexual orientation and earnings: A meta-analysis 2012–2020. *Journal of Population Economics* (Vol. 35, Issue 2). Springer Berlin Heidelberg. https://doi.org/10.1007/s00148-021-00862-1

Fitzgerald, K. G., & Caro, F. G. (2014). An overview of age-friendly cities and communities around the world. *Journal of Aging and Social Policy, 26*(1–2), 1–18. https://doi.org/10.1080/08959420.2014.860786

Foster, L., & Walker, A. (2021). Active ageing across the life course: Towards a comprehensive approach to prevention. *BioMed Research International, 2021*, 1. https://doi.org/10.1155/2021/6650414

Gandasari, D., & Dwidienawati, D. (2020). Content analysis of social and economic issues in Indonesia during the COVID-19 pandemic. *Heliyon, 6*(11), e05599. https://doi.org/10.1016/j.heliyon.2020.e05599

Ganz, F., Torralba, R., & Oliveira, D. V. (2020). Impacto de la inmovilización social por COVID-19 en la salud de personas adultas mayores: efectos físicos y mentales y recomendaciones. *The Journal of Nutrition, Health & Aging, 24*(9), 938–947. https://doi.org/10.0.3.239/s12603-020-1469-2.

Grande, G., Qiu, C., & Fratiglioni, L. (2020). Prevention of dementia in an ageing world: Evidence and biological rationale. *Ageing Research Reviews, 64*(February), 101045. https://doi.org/10.1016/j.arr.2020.101045

Havighurst, R. J. (1961). No title. *The Gerontologist, 1*(1), 8–13. https://doi.org/10.1093/geront/1.1.8

Heathcote, G. (2000). Autonomy, health and ageing: Transnational perspectives. *Health Education Research, 15*(1), 13–24. https://doi.org/10.1093/her/15.1.13

Jagannathan, R., Kapoor, M., & Schaumburg, E. (2013). Causes of the great recession of 2007–2009: The financial crisis was the symptom not the disease! *Journal of Financial Intermediation, 22*(1), 4–29. https://doi.org/10.1016/j.jfi.2012.06.002

Kalache, A., & Gatti, A. (2003). Active ageing: A policy framework. *Advances in Gerontology = Uspekhi Gerontologii / Rossiiskaia Akademiia Nauk, Gerontologicheskoe Obshchestvo, 11*, 7–18. https://doi.org/10.1080/tam.5.1.1.37

Keating, N. (2022). A research framework for the United Nations Decade of Healthy Ageing (2021–2030). *European Journal of Ageing, 19*(3), 775–787. https://doi.org/10.1007/s10433-021-00679-7

Khojasteh, D., Davani, E., Shamsipour, A., Haghani, M., & Glamore, W. (2022). Climate change and COVID-19: Interdisciplinary perspectives from two global crises. *Science of the Total Environment, 844*(June), 157142. https://doi.org/10.1016/j.scitotenv.2022.157142

Khosravi, P., Rezvani, A., & Wiewiora, A. (2016). The impact of technology on older adults' social isolation. *Computers in Human Behavior, 63*, 594–603. https://doi.org/10.1016/j.chb.2016.05.092

Klaming, L., Robbemond, L., Lemmens, P., & Hart de Ruijter, E. (2023). Digital compensatory cognitive training for older adults with memory complaints. *Activities, Adaptation and Aging, 47*(1), 10–39. https://doi.org/10.1080/01924788.2022.2044989

Kohrt, B. A., Hruschka, D. J., Worthman, C. M., Kunz, R. D., Baldwin, J. L., Upadhaya, N., Acharya, N. R., Koirala, S., Thapa, S. B., Tol, W. A., Jordans, M. J. D., Robkin, N., Dev Sharma, V., & Nepal, M. K. (2012). Political violence and mental health in Nepal: Prospective study. *British Journal of Psychiatry, 201*(4), 268–275. https://doi.org/10.1192/bjp.bp.111.096222

Krishnaswami, A., Beavers, C., Dorsch, M. P., Dodson, J. A., Masterson Creber, R., Kitsiou, S., Goyal, P., Maurer, M. S., Wenger, N. K., Croy, D. S., Alexander, K. P., Batsis, J. A., Turakhia, M. P., Forman, D. E., Bernacki, G. M., Kirkpatrick, J. N., Orr, N. M., Peterson, E. D., Rich, M. W., et al. (2020). Gerotechnology for older adults with cardiovascular diseases: JACC state-of-the-art review. *Journal of the American College of Cardiology, 76*(22), 2650–2670. https://doi.org/10.1016/j.jacc.2020.09.606

Ku, I., & Kim, C. O. (2020). Decomposition analyses of the trend in poverty among older adults: The case of South Korea. *Journals of Gerontology – Series B Psychological Sciences and Social Sciences, 75*(3), 684–693. https://doi.org/10.1093/geronb/gby047

Lagiewka, K. (2012). European innovation partnership on active and healthy ageing: Triggers of setting the headline target of 2 additional healthy life years at birth at EU average by 2020. *Archives of Public Health, 70*(1), 1–8. https://doi.org/10.1186/0778-7367-70-23

Li, M., Li, C., & Zhang, M. (2018). Exploring the spatial spillover effects of industrialization and urbanization factors on pollutants emissions in China's Huang-Huai-Hai region. *Journal of Cleaner Production, 195*, 154–162. https://doi.org/10.1016/j.jclepro.2018.05.186

Liffiton, J. A., Horton, S., Baker, J., & Weir, P. L. (2012). Successful aging: How does physical activity influence engagement with life? *European Review of Aging and Physical Activity, 9*(2), 103–108. https://doi.org/10.1007/s11556-012-0098-0

Liu, L., Stroulia, E., Nikolaidis, I., Miguel-Cruz, A., & Rios Rincon, A. (2016). Smart homes and home health monitoring technologies for older adults: A systematic review. *International Journal of Medical Informatics, 91*, 44–59. https://doi.org/10.1016/j.ijmedinf.2016.04.007

Löckenhoff, C. E., & Carstensen, L. L. (2004). Socioemotionol selectivity theory, aging, and health: The increasingly delicate balance between regulating emotions and making tough choices. *Journal of Personality, 72*(6), 1395–1424. https://doi.org/10.1111/j.1467-6494.2004.00301.x

Lopez, A. L., Macasaet, L. Y., Ylade, M., Tayag, E. A., & Ali, M. (2015). Epidemiology of cholera in The Philippines. *PLoS Neglected Tropical Diseases, 9*(1), e3440. https://doi.org/10.1371/journal.pntd.0003440

Lowen, T., Davern, M. T., Mavoa, S., & Brasher, K. (2015). Age-friendly cities and communities: Access to services for older people. *Australian Planner, 52*(4), 255–265. https://doi.org/10.1080/07293682.2015.1047874

Lunenfeld, B., & Stratton, P. (2013). The clinical consequences of an ageing world and preventive strategies. *Best Practice and Research: Clinical Obstetrics and Gynaecology, 27*(5), 643–659. https://doi.org/10.1016/j.bpobgyn.2013.02.005

Malik, S., Lee, D. C., Doran, K. M., Grudzen, C. R., Worthing, J., Portelli, I., Goldfrank, L. R., & Smith, S. W. (2018). Vulnerability of older adults in disasters: Emergency department utilization by geriatric patients after hurricane Sandy. *Disaster Medicine and Public Health Preparedness, 12*(2), 184–193. https://doi.org/10.1017/dmp.2017.44

Marmot, M., Allen, J., Bell, R., Bloomer, E., & Goldblatt, P. (2012). WHO European review of social determinants of health and the health divide. *The Lancet, 380*(9846), 1011–1029. https://doi.org/10.1016/S0140-6736(12)61228-8

Menassa, M., Stronks, K., Khatmi, F., Roa Díaz, Z. M., Espinola, O. P., Gamba, M., Itodo, O. A., Buttia, C., Wehrli, F., Minder, B., Velarde, M. R., & Franco, O. H. (2023). Concepts and definitions of healthy ageing: A systematic review and synthesis of theoretical models. *eClinicalMedicine, 56*, 101821. https://doi.org/10.1016/j.eclinm.2022.101821

Merkel, S., & Kucharski, A. (2019). Participatory Design in Gerontechnology: A systematic literature review. *Gerontologist, 59*(1), E16–E25. https://doi.org/10.1093/geront/gny034

Muszyńska, M. M., & Rau, R. (2012). The old-age healthy dependency ratio in Europe. *Journal of Population Ageing, 5*(3), 151–162. https://doi.org/10.1007/s12062-012-9068-6

Ntengwe, F. W. (2006). Pollutant loads and water quality in streams of heavily populated and industrialised towns. *Physics and Chemistry of the Earth, 31*(15–16), 832–839. https://doi.org/10.1016/j.pce.2006.08.025

Okely, A. D., Kontsevaya, A., Ng, J., & Abdeta, C. (2021). 2020 WHO guidelines on physical activity and sedentary behavior. *Sports Medicine and Health Science, 3*(2), 115–118. https://doi.org/10.1016/j.smhs.2021.05.001

Oleksiyenko, A., Terepyshchyi, S., Gomilko, O., & Svyrydenko, D. (2021). 'What do you mean, you are a refugee in your own country?': Displaced scholars and identities in embattled Ukraine. *European Journal of Higher Education, 11*(2), 101–118. https://doi.org/10.1080/21568235.2020.1777446

Olivera, J., & Tournier, I. (2016). Successful ageing and multi-dimensional poverty: The case of Peru. *Ageing and Society, 36*(8), 1690–1714. https://doi.org/10.1017/S0144686X15000665

OMS. (2015). Physical activity strategy for the WHO European region 2016–2025. *Organización Mundial de La Salud, 1*.

Peace, S., Holland, C., & Kellaher, L. (2011). "Option recognition" in later life: Variations in ageing in place. *Ageing and Society, 31*(5), 734–757. https://doi.org/10.1017/S0144686X10001157

Peel, N., Bartlett, H., & McClure, R. (2004). Healthy ageing: How is it defined and measured? *Australasian Journal on Ageing, 23*(3), 115–119. https://doi.org/10.1111/j.1741-6612.2004.00035.x

Polack, S., Scherer, N., Yonso, H., Volkan, S., Pivato, I., Shaikhani, A., Boggs, D., Beck, A. H., Atijosan-Ayodele, O., Deniz, G., Örücü, A., Akıncı, İ., Hameed, S., Acarturk, C., & Patterson, A. (2021). Disability among Syrian refugees living in Sultanbeyli, Istanbul: Results from a population-based survey. *PLoS One, 16*(November), 1–15. https://doi.org/10.1371/journal.pone.0259249

Pruchno, R. (2019). Technology and aging: An evolving partnership. In *Gerontologist* (Vol. 59, Issue 1, pp. 1–5). Gerontological Society of America. https://doi.org/10.1093/geront/gny153

Qin, X., Godil, D. I., Khan, M. K., Sarwat, S., Alam, S., & Janjua, L. (2022). Investigating the effects of COVID-19 and public health expenditure on global supply chain operations: An empirical study. *Operations Management Research, 15*(1–2), 195–207. https://doi.org/10.1007/s12063-020-00177-6

Quinn, V. J. M., Dhabalia, T. J., Roslycky, L. L., Wilson, V. J. M., Hansen, J. C., Hulchiy, O., Golubovskaya, O., Buriachyk, M., Vadim, K., Zauralskyy, R., Vyrva, O., Stepanskyi, D., Ivanovitch, P. S., Mironenko, A., Shportko, V., & McElligott, J. E. (2022). COVID-19 at war: The joint forces operation in Ukraine. *Disaster Medicine and Public Health Preparedness, 16*(5), 1753–1760. https://doi.org/10.1017/dmp.2021.88

Robbins, T. D., Lim Choi Keung, S. N., & Arvanitis, T. N. (2018). E-health for active ageing; A systematic review. In *Maturitas* (Vol. 114, pp. 34–40). Elsevier Ireland Ltd.. https://doi.org/10.1016/j.maturitas.2018.05.008

Rodeschini, G. (2011). Gerotechnology: A new kind of care for aging? An analysis of the relationship between older people and technology. *Nursing and Health Sciences, 13*(4), 521–528. https://doi.org/10.1111/j.1442-2018.2011.00634.x

Rodrigues, M., & Franco, M. (2021). The role of citizens and transformation of energy, water, and waste infrastructure for an intelligent, sustainable environment in cities. *Smart and Sustainable Built Environment, 12*(2), 385–406. https://doi.org/10.1108/SASBE-06-2021-0094

Rudnicka, E., Napierała, P., Podfigurna, A., Męczekalski, B., Smolarczyk, R., & Grymowicz, M. (2020). The World Health Organization (WHO) approach to healthy ageing. *Maturitas, 139*(February), 6–11. https://doi.org/10.1016/j.maturitas.2020.05.018

Sghaier, S., Asiamah, N., Danquah, E., Opuni, F. F., & Hatsu, S. (2022). Information technology ability mediates the association between older adults' subjective age and social activity: A STROBE-compliant cross-sectional analysis. *Archives of Gerontology and Geriatrics, 103*(August), 104790. https://doi.org/10.1016/j.archger.2022.104790

Stern, Y., Chételat, G., Habeck, C., Arenaza-Urquijo, E. M., Vemuri, P., Estanga, A., Bartrés-Faz, D., Cantillon, M., Clouston, S. A. P., Elman, J. A., Gold, B. T., Jones, R., Kempermann, G., Lim, Y. Y., van Loenhoud, A., Martínez-Lage, P., Morbelli, S., Okonkwo, O., Ossenkoppele, R., et al. (2019). Mechanisms underlying resilience in ageing. *Nature Reviews Neuroscience, 20*(4), 246. https://doi.org/10.1038/s41583-019-0138-0

Strong, J., Varady, C., Chahda, N., Doocy, S., & Burnham, G. (2015). Health status and health needs of older refugees from Syria in Lebanon. *Conflict and Health, 9*(1), 1–10. https://doi.org/10.1186/s13031-014-0029-y

Stuchul, D. L., Prakash, M. S., & Esteva, G. (2021). From fear to hope: Learning from BIPOC in hard times—Covid 19, climate collapse & racial violence. *International Journal of Lifelong Education, 40*(4), 415–426. https://doi.org/10.1080/02601370.2021.1936234

Tao, Y., Zhang, W., Gou, Z., Jiang, B., & Qi, Y. (2021). Planning walkable neighborhoods for "aging in place": Lessons from five aging-friendly districts in Singapore. *Sustainability, 13*(4), 1742.

Tosun, O. K., & Eshraghi, A. (2022). Corporate decisions in times of war: Evidence from the Russia-Ukraine conflict. *Finance Research Letters, 48*(April), 102920. https://doi.org/10.1016/j.frl.2022.102920

Van Der Cammen, T. J. M., Albayrak, A., VoûTe, E., & Molenbroek, J. F. M. (2017). New horizons in design for autonomous ageing. *Age and Ageing, 46*(1), 11–17. https://doi.org/10.1093/ageing/afw181

World Health Organisation. (2021). *WHO-EURO-2021-3409-43168-60449-eng*. https://www.who.int/europe/publications/i/item/WHO-EURO-2021-3409-43168-60449

Yusif, S., Soar, J., & Hafeez-Baig, A. (2016). Older people, assistive technologies, and the barriers to adoption: A systematic review. *International Journal of Medical Informatics, 94*, 112–116. https://doi.org/10.1016/j.ijmedinf.2016.07.004

Index

N. Asiamah et al. (eds.), *Sustainable Neighbourhoods for Ageing in Place*,
https://doi.org/10.1007/978-3-031-41594-4

The manufacturer's authorised representative in the EU is Springer Nature Customer Service Centre GmbH, Europaplatz 3, 69115 Heidelberg, Germany. If you have any concerns regarding our products, please contact ProductSafety@springernature.com

Printed and bound by CPI Group (UK) Ltd, Croydon, CR0 4YY
01/10/2026
02225892-0002